Engineering a cathedral

Engineering a cathedral

Proceedings of the conference *Engineering a cathedral* held at Durham Cathedral on 9–11 September 1993 as part of the 900th anniversary celebrations of Durham Cathedral

Edited by Rev. Dr Michael Jackson

Thomas Telford, London

A CIP catalogue record for this book is available from the British Library.

ISBN 0-7277-1684-0

First published 1993

© Authors, 1993, unless otherwise stated.

Papers or other contributions and the statements made or the opinions expressed therein are published on the understanding that the author of the contribution is solely responsible for the opinions expressed in it and that its publication does not necessarily imply that such statements and/or opinions are or reflect the views or opinions of the organizers or publishers.

All rights, including translation, reserved. Except for fair copying, no part of this publication may be reproduced, stored in a retrieval system or transmitted in any form or by any means electronic, mechanical, photocopying, recording or otherwise, without the prior written permission of the Publications Manager, Publications Division, Thomas Telford Services Ltd, Thomas Telford House, 1 Heron Quay, London E14 4JD.

Published on behalf of the organizers by Thomas Telford Services Ltd, Thomas Telford House, 1 Heron Quay, London E14 4JD.

Printed in Great Britain by Redwood Books, Trowbridge, Wiltshire.

Foreword

The spectacular impressiveness of Durham Cathedral is obvious at first glance, whether that glance comes from the railway line or from one of the hills as one approaches the city. As one walks around the city and sees views from the River and on Palace Green, the impression of weight, solidity and dignity is reinforced. Once inside, the effect is different but equally amazing, with touches of awesomeness and majesty.

I have now had the benefit and privilege of some nine years driving round the Diocese and County, and of participating in many services and other occasions in the Cathedral. Glimpses of the Cathedral from viewpoints all over the Diocese and in all sorts of lights, add to the sense of the dignity and solidity of it. Within the Cathedral, perhaps gazing around while listening to the Psalms so clearly articulated by the choir, or while enjoying some music at a concert, the building keeps on revealing arches and spaces and glimmerings of light which add to the abiding wonder of the place.

All this eventually raises in one's mind and spirit a set of wondering questions — Who planned this place? How could they, and did they, think it out and then bring it about? Why and how does it stand, and stand so well for so long? How are we to care appropriately for it and maintain it, both in its solidity (no stones are impermeable to time and weather and to the stresses and strains put upon them by the very nature of the building) and its use for worship, praise, teaching and wonder?

I find my wondering in these respects reminds me, as far as sheer building goes, of reflections which have gradually developed in me through a number of visits I have made over the years to the great Byzantine Church of Hagia Sophia in Instanbul. History has caused it to pass from being the great Christian Cathedral of Constantinople, as capital of the Byzantine Roman Empire, through being a Mosque to its present existence as a Museum. The building is a wonder which only dawns upon you as you reflect upon it and keep quiet within it — a wonder in size, in proportions, in visual and sensual effectiveness. Men conceived such a design and put up such a building. How was it done? The same wonder arises about the building of Durham Cathedral as an architectural and engineering achievement carried out by men who could plan, supervise, dress and locate stone, then labour hard, and doubtless at great physical cost and risk, to bring such a building to completion. As we quietly contemplate the majesty, the might and the continuity of Durham Cathedral, we have indeed much to wonder at.

I am therefore very glad indeed that as part of the 900th Anniversary celebrations, a conference has been set up which addresses many of these questions I wonder about, around the theme of 'Engineering a Cathedral'. It is good that experts in

FOREWORD

architecture, history and building engineering have offered the papers which make the chapters of this book so that our wonder can be informed and enlightened. As they inform us about the achievements of the past, and about the needs and possibilities of maintenance for the future, they make no small contribution to our set of celebrations for the 900th Anniversary of the foundation of this great building. These papers also give us information which helps us to appreciate how, in practice, men turned a vision of the worship of God into an architectural and visual wonder. There is help and encouragement also in the task and privilege of maintaining the building; both in its wonder of stone and in its wonder as a living place of faith and worship, wisdom and comfort. So the Cathedral can continue to serve as it has done for 900 years by standing on its hill, speaking not of its own permanence, for that has its limits, but of the abiding majesty, presence and promise of God.

David Dunelm

Contents

Durham Cathedral 1093–1193 sources and history. D. ROLLASON	1
The building of the cathedral. C. DEVONSHIRE and R. WILKIE	16
An appraisal of the topology and loading of a section of the vaulted roof of Durham Cathedral. D. M. LILLEY	32
The Romanesque high vaults of Durham Cathedral. M. THURLBY	43
The purpose of the rib in the Romanesque vaults of Durham Cathedral. M. THURLBY	64
Durham Cathedral tower vibrations during bell-ringing. J. M. WILSON and A. R. SELBY	77
Cathedral engineering for novices. J. WELFORD	101
The geological setting of Durham Cathedral. G. A. L. JOHNSON	109
Deterioration of the exterior stone at Durham Cathedral. P. B. ATTEWELL	120
Design principles of early medieval architecture as exemplified at Durham Cathedral. PROFESSOR E. C. FERNIE	146
The unseen timber roofs of the choir, north transept, 1-2 The College and the dorter of Durham Cathedral. J. W. BULL	157
The roof of the monks' dormitory, Durham. J. HEYMAN	169
The maintenance problem. I. HUME	180
Maintenance management. S. D. STEVENS	188

Durham Cathedral 1093–1193 sources and history

D. ROLLASON, MA, PhD, FSA, FRHS, Senior Lecturer in History, University of Durham

SYNOPSIS. Although contemporary writings describing the construction of the Anglo-Norman cathedral of Durham are exiguous by modern standards, they are worth close study. The most directly relevant writings are here presented in English translation with historical commentary.

EARLY HISTORY
The construction of the Anglo-Norman cathedral of Durham was one of the greatest artistic and engineering achievements of the Middle Ages. The building itself is unquestionably the most valuable historical source for understanding how it was erected and why it was designed in the particular ways it was. Nevertheless there are nearly contemporary written sources which cast light on the development of the building and its site, and they deserve close study, even though they are inevitably more exiguous and more obscure than records relating to a modern building on a scale comparable to Durham cathedral might be expected to be. For it is important to emphasize that there are no historical sources of which the prime purpose was to describe the progress of the work. Most of the information at our disposal was provided as incidental detail to accounts of the bishops, stories of the miracles of St Cuthbert, and so on. The aim of this contribution is to set out the main relevant extracts from the near contemporary sources, and to set them as far as possible in the context of the history of the cathedral itself. The original Latin versions of all the texts discussed here will be found in the collection of Otto Lehmann-Brockhaus (ref. 1), and there is a full historical and documentary commentary by Martin Snape (ref. 2).

Any account of Durham cathedral's early history must begin with the work which has generally been known as the *History of the Church of Durham*, the original title of which was *Tract on the Origin and Progress of this the Church of Durham* (the 'progress' in question is that of the bishops, saints and community of the church rather than primarily of its buildings). Evidence from the content of this work shows that it must have been written between 1104 and 1109, possibly before 1107, and it is in fact preserved in two manuscripts of the very early twelfth century, one in London and another (which must be very close to the author's own work) in the University Library, Durham, where it is Cosin MS V.II.6. The author was probably a Durham monk called Symeon, who is named as author in another manuscript of the work, which dates from the later twelfth century and is preserved in Cambridge University Library (MS Ff.i.27) (refs 3, 4, 5, 6).

The *Tract*, as we shall call it, was clearly intended to be read in Durham itself and in the Cosin manuscript there is a long list of the monks of Durham prefaced to the main text. Despite its title the *Tract* begins with the foundation of the monastery of Lindisfarne in 635, describes the career of St Cuthbert (d.687), and then follows the fortunes of the Lindisfarne community when it fled from the Vikings in 875 and settled for over a century at Chester-le-Street (883-995). Down to that point, Symeon drew on a number of earlier sources which still survive, such as Bede's *Ecclesiastical History of the English People* and his *Life of St Cuthbert*, but his *Tract* is the earliest source for the removal of the former Lindisfarne community from Chester-le-Street to Durham in 995. The story he tells is well known: how in the face of fresh Viking threats the community, led by Bishop Ealdhun (990-1018), moved to Ripon, how as they were making the return journey the coffin of St Cuthbert became too heavy to lift, and how following prayers and vigil the saint's wish to rest on the peninsula at Durham was divinely revealed (ref. 7). What concerns us here is exactly what Symeon has to say about the site chosen on the peninsula. His words in English translation are given below, cited by book and chapter as in Arnold's edition (ref. 3; III.2):

> *The whole people who accompanied the body of the most holy father Cuthbert to Durham found there a place which, although it possessed natural defences, was not easily habitable because it was completely covered on all sides by very dense forest. Only in the middle was there a piece of level ground and this was not large. They had been accustomed to cultivate this by ploughing and sowing, but later Bishop Ealdhun built a stone church of some size on it, as will appear further on in our account. So the aforesaid bishop, with the help of all the people and the assistance of Uhtred, earl of the Northumbrians, cut down and uprooted the whole forest and soon made the place habitable. Later, a multitude of people from the whole area between the River Coquet and the River Tees readily came to help not only with this task but also afterwards with the construction of the church, and they persevered devotedly until it was finished.*

The passage is not easy to understand in the original Latin, and it is by no means clear whether it means that the community itself was cultivating the central area of the peninsula or that it had already been cultivated before their arrival (ref. 8). Nevertheless we catch some glimpse of the character of the place with its natural fortifications, and we see too the participation of Uhtred, the earl of Bamburgh and the secular ruler of Northumbria. It is clear that even from this period the peninsula was a military fortress as well as an ecclesiastical site. Indeed we possess a late eleventh-century text called *The Siege of Durham* which describes the repulse of a Scottish army, apparently in 1006, and particularly the impaling of the heads of defeated Scots on the walls of Durham (ref. 9).

In describing the arrival of the former Chester-le-Street community at Durham, Symeon writes that the body of St Cuthbert was placed temporarily in 'a little church of branches' (III.1). Following on from his account of the clearing of the peninsula he then explains the sequence of church construction on the site (III.2):

> *When the forest had been eradicated and dwellings assigned to each by lot, Bishop Ealdhun, who was burning with love for Christ and St Cuthbert, began to build a church of noble workmanship and by no means small in scale, and to the completion of this he devoted all his efforts. Meanwhile the holy body was translated from that little church which we mentioned above [i.e. the church of branches] into another which was called the White Church, and there it remained for three years while the larger church was being built.*

The implication of Symeon's words would seem to be that there were in all three churches: the temporary church of branches to house the body of the saint on its arrival; the 'White Church'; and the larger church referred to at the end and which the first bishop of Durham, Ealdhun (d.1018), is said later in the account to have built. There are, however, some problems with this interpretation deriving from the confusion which arose at Durham from the later twelfth century onwards as to whether Bishop Ealdhun's church was in fact the 'White Church'. In any case, virtually nothing is known about these churches, even their sites, although there is some ground for supposing that the last of them occupied a site parallel to the present cathedral in what is now the cloisters (ref. 10). Symeon tells us that at the time of his death in 1018 Bishop Ealdhun had finished this church apart a western tower which his successor Edmund (c.1020-c.1040) completed (III.5), and we do catch a glimpse of this tower later on in the *Tract*. After telling us something of the life of the community at Durham in the eleventh century - of which he did not altogether approve since the members of the community were clerks rather than monks and appear to have been married - Symeon describes the Norman impact on Durham and he devotes a detailed chapter to the arrival of the first Norman earl Robert Cumin in 1069. Before reaching Durham this man allowed his soldiers to ravage the countryside and to kill the church's peasants so that a dreadful fate awaited him in Durham. As he was spending the night (in the bishop's house, as we learn from another twelfth-century northern source, the *History of the Kings* (ref. 11)) Northumbrian insurgents entered the city.

> *They set fire to the house, and tried to burn it down together with those who were inside. As balls of fire flew up high to a great height, it seemed that the west tower which stood nearby was bound to be burned. So the people knelt down and beseeched St Cuthbert that he should preserve his church unharmed from the flames, and at once a wind sprang up from the east and blew the balls of flame away from the church, and repelled all danger far from it.*

The information to be derived from this is exiguous indeed, but we do see a west tower to the church and we see also the bishop's house near enough to it for the fire to pose a threat. There is in fact a more detailed description in the *Little Book about the Wonderful Miracles of the Blessed Cuthbert which were Performed in Recent Times*, probably completed in 1174 by the Durham writer Reginald (ref. 12). It is not clear what the source of Reginald's information was, and his work is marred by the confusion referred to above over the identity of the 'White Church', but for what it is worth he writes (ref. 13):

> *There were in the White Church in which [St Cuthbert] had first rested, two stone towers, as those who saw them have told us, standing high into the air, the one containing the quire, the other standing at the west end of the church, which were of wonderful size. They carried brazen pinnacles set up on top which aroused the amazement of all and a great deal of admiration.*

It has been suggested on the basis of this that Bishop Ealdhun's church, assuming that to be the church to which Reginald was referring, was like the surviving late Saxon churches at Dover and, nearer at hand, at Norton near Stockton (ref. 14). The description is thus interpreted to mean that there was a western as well as an eastern tower and with the eastern tower standing over the crossing of a cruciform building (ref. 15).

BISHOP WALCHER (1071-80)

Returning to Symeon's narrative, this then moves on to the arrival in the north in 1073-4 of three monks from the Worcestershire monasteries of Pershore and Evesham. These men, led by an Englishmen called Aldwine, had read in Bede's *Ecclesiastical History of the English People* of the former monasteries in Northumbria which no longer existed, presumably in part at least because of the activities of the Vikings. They settled first at Newcastle, but then the first bishop of Durham appointed by the Norman kings, Walcher (1071-80), moved them to Bede's old monastery at Jarrow which was then in a ruined condition. They re-established monastic life not only there but also at Monkwearmouth, which Walcher also gave them, and at Whitby which had likewise figured as an important monastery in Bede's work (ref. 16).

Walcher was himself a clerk, that is he was supposed to lead his life according to a rule less strict than that of the monks, but according to Symeon (III.22):

> *He had decided that, if the period of his life were to be of longer duration, he too would become a monk, and would establish a dwelling of monks around the holy body of the blessed Cuthbert. For which reason he laid the foundations and began to construct buildings for monks where they now are at Durham. But alas! he was prevented by death, and was unable to complete what he had decided upon.*

It is not certain how far Walcher's abortive project influenced the subsequent development of the cathedral site. A summary of Symeon's work prefaced to it speaks of the buildings for the monks as being 'around the walls of the church' (ref. 17), and a foundation recovered in the present cloisters may be a relic of them, as also may be the early masonry visible in the undercroft of the refectory, the undercroft of the original dormitory on the east side of the cloisters, the entry in the south-east corner of the cloisters, and the lower part of the east wall of the western range (ref. 14).

Walcher's career in Durham is perhaps most interesting for the light it casts on the violent conditions of the first years of Norman rule in the north. Problematic as the bishop's contribution to the Norman cathedral's earliest phases may be, the Norman castle was built for him as is made clear by the *History of the Kings*, although the grammar of the Latin makes it problematic as to whether it was built by Earl Waltheof of Northumbria or King William himself. The entry refers to the year 1072 (ref. 18):

> After the overthrow of Cospatrick, Waltheof was raised to the earldom...At that time, that is when the king had returned from Scotland, he built in Durham a castle where the bishop could keep himself and his men safe from the incursions of attackers.

How necessary this was is shown by the savage fate the bishop eventually suffered. Although Symeon describes the events in outline, the fullest source is the early twelfth-century chronicle known as the *Chronicle of Florence of Worcester*. According to this Walcher's kinsman Gilbert went at night to the house of a Northumbrian called Ligulf, a scion of the house of the earls of Bamburgh, and massacred him and his whole family. This in effect initiated a blood-feud. Walcher failed to dissociate himself from Gilbert's action or even to remove him from his entourage, so the wrath of the house of Bamburgh was directed against the bishop. He eventually agreed to meet their representatives at Gateshead, but the meeting was apparently a trap. When Walcher withdrew into a small church, the Northumbrians attacked his knights and set fire to the church. When the bishop was at last driven out by smoke and heat, they massacred him (ref. 19).

BISHOP WILLIAM OF ST CALAIS (1080-96)

It is not certain how far the clerks of the community of Durham were implicated in this killing, but it is certain that they were closely associated with the house of Bamburgh so they may at least have been 'accessories after the fact' (ref. 20). In any case, the next Norman bishop, William of St Calais (1080-96), was himself a monk and would have thought in terms of establishing a monastery at his cathedral church, especially as he was influenced by Lanfranc, archbishop of Canterbury (1070-89), who had established such a community on a firm footing at his own church (ref. 21). At any rate, Symeon represents him first as consulting the elders of his church and discovering from them and from the work of Bede that there had been monks at Lindisfarne with St Cuthbert, and then as seeking the consent of Archbishop Lanfranc, King William I, Queen Mathilda, and even Pope Gregory VII. Then he moved Aldwine and his monks from Jarrow and Monkwearmouth and brought them to Durham to form the community of the cathedral. As for the clerks who had previously served the church (IV.3):

> The bishop ordered that those who had previously dwelt in the church, and who had been canons only by name since they in no way followed the rule of canons, should henceforth lead their lives with the monks and according to the monastic vocation if they wished to remain in the church. But they preferred to leave the church rather

> than to enter on those terms, except one who was their dean and who was persuaded with difficulty by his son who was a monk that he also should become a monk.

This was in 1083 and it might have been expected that William of St Calais would embark at once on building works, partly to complete Walcher's construction work to accommodate the monks, partly to emulate the great building projects of other Norman bishops. In fact, William had at first no leisure for this since he was closely involved in secular affairs. Almost certainly he was one of the commissioners responsible for the drawing up of *Domesday Book* (1086), and in William Rufus's reign (1087-1100) he became embroiled in the revolt of Bishop Odo of Bayeux and the civil war between the king and his brother Robert Curthose, duke of Normandy. He seems ill-advisedly to have taken the side of the latter and was exiled by William Rufus to Normandy from 1087 until 1090 (ref. 22, 23, 24).

Symeon tells us, however, that during his absence 'the monks built the refectory as it appears today', and that eventually the bishop was able to make his peace with the king and he was restored to his see, bringing back with him 'many sacred altar vessels of gold and silver, various ornaments, and also many books'. Then the real work of construction began (IV.8).

> Not long afterwards, in the ninety-eighth year since it had been founded by Ealdhun [1092], he ordered the church to be demolished and after he had laid the foundations in the following year, he began to construct another on a nobler and grander scale.

This passage is a little puzzling, since it is hard to see where services would be held during the construction of the new church if the old one had been completely demolished. Nor is it clear where the remains of St Cuthbert and the church's other saints were housed during the building work. There must obviously have been a church still in working order. One possibility is that the first of the churches built by Bishop Ealdhun in the late tenth century was still available (see above); another is that Symeon is in error - or at least that his words are ambiguous - and that the Saxon cathedral was left standing alongside the new Norman one - that is on the site of the present cloister - at least until Cuthbert's body was translated into its shrine in the new church in 1104. That is the implication of the accounts of that translation described below, and it would also be in line with the architectural history of other large churches, notably Winchester and Wells (ref. 25, 26). At all events, Symeon goes on to describe the actual inception of the work:

> This was begun on Thursday 11 August in the year 1093 of the Lord's Incarnation, the thirteenth of William's time as bishop, the eleventh since he had brought together the monks at Durham. On that day the bishop and Prior Turgot, who was second in authority to him in the church, with the other brothers laid the first stones in the foundations. Shortly before, that is on Friday 29 July, the bishop and prior after saying prayers with the brothers and giving their blessing had begun to dig the foundations. While the monks were

> *responsible for building the monastic buildings, the bishop carried out the work on the church at his own expense.*

This was obviously a major event. There is no reason to think that the monks - let alone the bishop and the prior - were physically responsible for the building works and we must suppose that the digging of the foundations referred to in the text was of a ceremonial character, comparable to the modern ceremony of 'laying the foundation stone'. Indeed the *History of the Kings* records that Malcolm, king of Scots, was also present, which is surprising in view of the fact that later in the year he was to be killed while invading Northumbria (ref. 27):

> *The new church of Durham was begun on Thursday 11 August, with Bishop William, Malcolm, king of Scots, and Prior Turgot laying the first stones in the foundations.*

Aside from a reference to the burial of William himself in the chapter-house at Durham (IV.10), this is all Symeon has to say about the building of Durham cathedral and he closes his *Tract* with William's death in 1096. The only other source to cast any light on the early progress of the Norman cathedral is a group of stories describing miracles worked in association with St Cuthbert, and apparently written in Durham around the year 1100 (ref. 28). One of these, which may in fact be earlier than the others as will appear, has some bearing on the construction of the cathedral, for it provides a glimpse of the transportation of timber for the building work (ref. 29):

> *A year has not yet passed since a baulk of wood was being transported from a distance for use in the construction of the church which is now being built of such beautiful workmanship in his [St Cuthbert's] honour. This is said to have been of such size that no less than eight oxen pulling with all their strength had to draw it. It was brought near to the gate of the city, where the tired oxen were allowed to breathe a little, so that having regained their strength they might more easily exert themselves against the slope.*

The real point of the story is then reached. It appears that a crowd of boys was playing and fooling about, and that when the baulk moved off again a child among them seemed to be crushed under it. The other boys called back the drivers who had not noticed the accident, and with great difficulty sixteen men raised the baulk and released the child - who, thanks to St Cuthbert, was found to be miraculously unharmed. The story cited above illustrates the difficulty of using this type of source material. There is, as might be expected, no precision as to the size of the baulk involved, and we are left with a suspicion that even the general account of its size may have been exaggerated to highlight the efficacy of Cuthbert's miraculous intervention in saving the child from injury.

BISHOP RANULF FLAMBARD (1099-1128)

After William of St Calais's death on 1 January 1096, the king kept the see of Durham vacant for over three years so that he could enjoy the revenues from it. Then at Whitsun 1099 he gave it to his close adviser and favourite Ranulf Flambard who had a bad reputation among his contemporaries as a tax-gatherer and an exploiter of churches. When William Rufus was killed on 2 August 1100, Ranulf was almost immediately imprisoned in the Tower of London by his successor Henry I and had to escape to Normandy by the aid of a rope smuggled to him in a wine-stoup. Eventually Ranulf became reconciled to the king and was able to become a surprisingly effective bishop of Durham, especially as regards building works (ref. 30). Our principal source for this is an anonymous continuation of Symeon's *Tract*, the earliest version of which is preserved in the Cosin manuscript referred to above. The hand of the scribe seems to be contemporary with Ranulf and, since it finishes and a new scribe takes over just after an account of Ranulf's death, it seems reasonable to assume that the first continuator was writing during or immediately after Ranulf's lifetime (ref. 31). Much of the continuation is concerned with Ranulf's earlier career, especially an exciting escape he had from would-be enemies at sea, but there is a substantial passage devoted to construction work at Durham (ref. 32):

> *Towards the building of the church he acted at times more assiduously and at others with more remissness, depending on whether offerings made at the altar and dues from the cemetery were available or had ceased to flow in. For with the income from these he had erected the walls of the nave up to its vault. Now, his predecessor, who began the construction, had laid it down as a decision that the bishop should pay for the church from his own resources, and the monks should pay for the monastic buildings out of what they collected from the church. With his death, however, this arrangement collapsed. For the monks neglected the monastic buildings and concentrated on the construction of the church, which Ranulf consequently found built as far as the nave. He also added to the ornaments of the church dorsals, palls, copes, chasubles, tunics and dalmatics; and he extended the cramped precinct (curia) of the monks by a considerable distance widthways and lengthways.*

The arrangements made by William of St Calais for the financing of the building cast some light on the organization of the construction, and it is interesting that the monks should have been so anxious to proceed with the church that they shouldered what had at the beginning been the bishop's responsibility. Light is also cast on the sequence of building. As might be expected, the east end was built first, and this was largely complete by Ranulf's time. It is not clear how much of the transept is meant by the phrase 'up to the nave', but confirmation of the fact that the east end was complete by 1104 is provided by three other early sources (ref. 33). These all concern the translation of the body of St Cuthbert into the new Norman church in 1104 and they are: another miracle story from the group referred to above and describing the translation in considerable detail (ref. 34, 35); an even more detailed account of that translation in the Reginald's *Little Book about the Wonderful Miracles of the Blessed Cuthbert* (ref. 36, 37); and a brief account of the translation with miracle stories by William of Malmesbury, who in 1125 wrote a work called the *Deeds of the Bishops*, summarizing the histories of all the English sees (ref. 38). The details of the translation itself need not concern us here except to note that its date is clearly fixed (29 August 1104) and that the fact that it took place then suggests that the eastern arm of the cathedral was complete so as to receive the saint's body and to permit the monks to hold their services in the church. One of the stories told by William of Malmesbury confirms this and casts some light on structural matters. It reads:

> *Everything in the new church was prepared for the translation of the body: the choir of the monks, the altar, the sepulchre. All that was needed was that the wooden material, which was holding up the new roof of the presbytery, should be carefully taken down. But you did not suffer, O most blessed one! the holy desire of your devotees to be prolonged, and you brought it all down in the dead of night. For who else should have been the author of such a deed? Alarmed by the sound the prior came running, fearing for the safety of the altar and the pavement, but not at all concerned about the wood. But you not only preserved intact the things which were feared for, but also the wood itself just as it had been erected. Justly therefore you are feared by your monks, so that none of them who is guilty of disobedience to the prior or conscious of having committed some grave offence dares to pass the night without having made confession of his fault.*

The last sentence shows again the difficulties of using this type of source material, since the emphasis is clearly on the miraculous powers of St Cuthbert and the necessity of monastic obedience. Nevertheless, the details of the story, although the use of Latin words is not entirely clear, do seem to relate to the wooden centring which was still supporting the vault of the 'presbytery', presumably the eastern arm above the shrine. The implication is that on the eve of the translation the vault had literally only just been completed, perhaps because it still needed such support before the mortar dried, perhaps because the builders had were behind with their schedule and there had been no time to remove it.

The passage from the anonymous continuation of Symeon's *Tract* cited above shows that Ranulf Flambard had built the nave 'up to the *testudo*' (possibly the roof-plate but more likely the vault) before his death in 1128, but it was evidently not complete because a passage in a further continuation of Symeon in the Cosin manuscript notes that while the bishopric was vacant for five years after Ranulf's death 'the nave of the church of Durham was brought to completion through the assiduity of the monks' (ref. 39).

The earlier continuation also describes Ranulf's treatment of what is now Palace Green (ref. 40):

> *Although the town was naturally fortified, he made it still stronger and more majestic with a wall. From the chancel of the church to the stronghold of the castle he constructed a wall of great length. The space between the church and the castle, which had been occupied by many dwellings, he made as flat and open as a field, so that there should be no risk to the church either from filth and rubbish, or from fires.*

Thus the open aspect of the cathedral towards the castle is the work of Flambard. Indeed the Latin may even imply that he was responsible for levelling Palace Green, a process which seems to have left some evidence in the archaeological record (ref. 41).

THE MIDDLE YEARS OF THE TWELFTH CENTURY (1128-52)
The middle years of the twelfth century were often troubled ones at Durham in part because of the effects of the civil war of King Stephen's reign. This was particularly so after the death of Bishop Geoffrey Rufus in 1141 when William Cumin, chancellor to the king of Scots, seized the see of Durham against the wishes of the monks. A long struggle ensued, which was itself embroiled in the civil war and was often violent with ravaging of the countryside and at one point the forcible expulsion of the monks from the cathedral, before in 1144 Cumin was induced to hand over his see to the properly elected bishop, William of Ste Barbe (1144-52) (ref. 42).

As might be expected, there was little real building during these years. The further continuation of Symeon reports that 'in the time of Bishop Geoffrey the chapter house of the monks was finished', and that the dormitory (presumably the original eastern dormitory) was completed under William of Ste Barbe (ref. 43). Some effort was evidently expended on the adornment of the church, for Reginald of Durham has an interesting account of this (ref. 44):

> *Roger, prior of Durham [?1138-49], a man distinguished by character and deeds, had grown great in his devotion to the glory of St Cuthbert, so that he decided to lay a pavement of marble in his church. For because he had heard that foreign churches in overseas countries were resplendent in the beauty of such work, he decided, for the glory of St Cuthbert, to adorn and beautify his church with such a scheme of construction. To this end he called together and consulted various devout people who were going for religious reasons to foreign countries, and he urged and entreated them to bring back for him pieces of marble for this work.*

The story goes on to describe how a certain Harpin, a knight of Thornley, made a pilgrimage to Rome and was able to bring back 'a modest quantity of marble which was nevertheless of a kind conspicuous for its outstanding quality'.

BISHOP HUGH OF LE PUISET (1153-95)
Real further progress in the enlargement and decoration of the church was made by the next bishop, Hugh of le Puiset (1153-95). He has been described as 'one of the most remarkable men of his time' (ref. 45, 46). The nephew of the great bishop of Winchester, Henry of Blois, he was deeply involved in national political life, but he was nevertheless an assiduous, effective, and princely bishop of Durham, responsible not only for much building work but also for the drawing up of the great survey of the bishopric's estates known as *Boldon Book*. The second continuation of Symeon says of him (ref. 47):

> He constructed many buildings in the bishopric, and not least in the city of his see where, having demolished old buildings, he built new and distinguished ones. To his cathedral church, in which the body of the most blessed Cuthbert rests, he gave many ornaments and he extended the church with distinguished work, and made it both longer and more glorious. He decorated the whole building with marble from afar, and increased the number of finely painted glass windows around the altars.

A more detailed account of le Puiset's work on the cathedral is contained in a further source, the *Book of Geoffrey, sacrist of Coldingham, concerning the condition of the church of Durham*. Coldingham was a monastic cell of Durham and had Geoffrey formerly been a monk of Durham. His work covered the period from 1152-1214, for much of which he could give a contemporary account. His description of le Puiset's first attempt at extending the cathedral is very remarkable (ref. 48):

> He approved of the efforts of his predecessors who had built the church and had brought to it many things for its adornment and its protection, and he strove diligently to emulate those who he knew to be men of probity. He therefore began to erect a new structure at the east end of the church. He had columns and bases of marble brought from overseas. When by the intervention of fate, there were admitted many masters, not without danger to themselves, and there were as many beginnings as there were masters, having incurred copious expenses for the workmen, and the walls hardly erected to any height at all, and at length split apart into cracks, manifest sign was given that this was not acceptable to God and his servant St Cuthbert.

Any remains of this abortive extension of the east end have been swept away by the construction of the Gothic Chapel of the Nine Altars in the middle years of the thirteenth century, but the account casts an interesting light on le Puiset's plans for the east end as well as on the fallibility and limitations of medieval architects. Geoffrey goes on:

> So he abandoned that work and began the one at the west end to which entry would be allowed to women so that, not having bodily access to the secret places of the saints, they should at least have some consolation from contemplating them.

This of course refers to the construction of the Gallilee Chapel and Geoffrey provides the earliest evidence for its use as a place to which women might be admitted. Geoffrey does not describe it as a Lady Chapel but it is so designated by a charter of the 1180s which shows that it was in existence and dedicated to St Mary by 1189 at the latest (ref. 49).

With the work of le Puiset the Anglo-Norman church of Durham reached its apogee, both in terms of its structure and in terms of its ornament, for a series of bishops had enriched it in this respect, and Geoffrey waxes lyrical about the treasures presented to it by Hugh, especially great lamps made of precious metals and crystal around the altars, and a wonderful shrine for the bones of Bede (ref. 50). As with the cathedral, so with the city. Le Puiset was responsible for the construction of Elvet Bridge and for the borough of Elvet. He undertook substantial building works in the castle, and he moved the site of Kepier Hospital, constituting it for a master and thirteen brethren under monastic rule (refs. 45, 51, 52). The century from 1193 to his death had indeed been a remarkable one for Durham and, although many changes were to made in the following years, the core of the city of Durham, with its cathedral, castle, fortified walls, and urban development was established.

REFERENCES

1. LEHMANN-BROCKHAUS O. Lateinische Schriftquellen zur Kunst im England, Wales und Schottland vom Jahre 901 bis zum Jahre 1307, vol. 1, pp.351-405. Prestel, Munich, 1955.

2. SNAPE M.G. Documentary Evidence for the Building of Durham Cathedral and its Monastic Buildings, in The British Archaeological Association Conference Transactions for the year 1977, III, Medieval Art and Architecture at Durham Cathedral, ed. N. Coldstream and P. Draper, pp.20-36. British Archaeological Association, Leeds, 1980. I am very grateful to Mr Snape for advice with this paper, although errors remain my own.

3. ARNOLD T. (ed.) Symeonis monachi Opera omnia, vol. 1, pp. ix-xxiii, and pp.3-135. Rolls Series, Longmans, London, 1882-5.

4. STEPHENSON J. (trans.) Simeon of Durham, A history of the church of Durham. Llanerch, Lampeter, 1988.

5. ROLLASON, D.W. (ed. and trans.) Libellus de exordio atque procursu istius, hoc est Dunhelmensis ecclesie. Oxford University Press, Oxford, forthcoming.

6. OFFLER H.S. Medieval Historians of Durham, pp. 6-8. University of Durham, Durham, 1958.

7. ROLLASON D.W. (ed.) Cuthbert, saint and patron, pp. 45-59. Dean and Chaper of Durham, Durham, 1987.

8. CARVER M.O.H. Early medieval Durham: the archaeological evidence, in Medieval Art and Architecture at Durham Cathedral (ref. 2), pp. 11-12.

9. HART C.R. The early charters of northern England and the north midlands, pp. 143-50

10. CAMBRIDGE E. et al. A new approach to church archaeology: dowsing, excavation and documentary work at Woodhorn, Ponteland and the pre-Norman cathedral at Durham. Archaeologia Aeliana, 1983, 5th series, vol. 11, 91-7.

11. ARNOLD T. (ed.) Symeonis monachi Opera omnia (ref. 3), vol. 2, pp. 186-7.

12. TUDOR V. The cult of St Cuthbert in the twelfth century: the evidence of Reginald of Durham, in St Cuthbert, his cult and his community, ed. G. Bonner, D. Rollason and C. Stancliffe, p. 449. Boydell and Brewer, Woodbridge, 1989.

13. RAINE J. (ed.) Reginaldi monachi Dunelmensis Libellus de admirandis beati Cuthberti virtutibus quae novellis patrarae sunt temporibus, p. 29. Surtees Society, Durham, 1835.

14. HOPE W.S. Notes on recent excavations in the cloister of Durham abbey. Proceedings of the Society of Antiquaries, 1901, vol. 22, 416-23.

15. CLAPHAM A.W. English romanesque architecture before the Conquest, p.88. Oxford University Press, Oxford, 1930.

16. KNOWLES D. The monastic order in England 940-1216, 2nd edition, pp. 165-8. Cambridge University Press, Cambridge, 1963.

17. ARNOLD T. (ed.) Symeonis Opera (ref. 3), vol. 1, p. 10.

18. ARNOLD T. (ed.) Symeonis Opera (ref. 3), vol. 2, pp. 199-200.

19. THORPE B. (ed.) Florentii Wigorniensis monachi, Chronicon ex chronicis, vol.2, pp. 13-16. English Historical Society, London, 1849.

20. ROLLASON D.W. Symeon of Durham and the community of Durham in the eleventh century, in England in the eleventh century: Proceedings of the 1990 Harlaxton Symposium, ed. C. Hicks, pp. 183-98. Paul Watkins, Stamford, 1992.

21. KNOWLES D. Monastic order (ref. 16), pp. 129-34.

22. STEPHEN, L. et al. Dictionary of national biography, vol. 3, cols. 990-3. Oxford, Oxford University Press, Oxford, 1921-2.

23. CHAPLAIS P. William of Saint-Calais and the Domesday Survey, in Domesday Studies, ed. J. C. Holt, pp. 65-77. Boydell and Brewer, Woodbridge, 1987.

24. OFFLER H.S. William of St Calais, first Norman bishop of Durham. Transactions of the Architectural and Archaeological Association of Durham and Northumberland, 1946-53, vol. 10, 258-79.

25. SNAPE (ref. 2), p. 21.

26. CAMBRIDGE (ref. 10), pp. 93-4.

27. ARNOLD T. Symeonis Opera (ref. 3), vol. 2, p. 220.

28. COLGRAVE B. The post-Bedan miracles and translations of St Cuthbert, in The early cultures of north-west Europe (H.M. Chadwick memorial studies), ed. C. Fox and B. Dickins, pp. 307-32. Cambridge University Press, Cambridge, 1950.

29. ARNOLD T. (ed.) Symeonis Opera (ref. 3), vol. 2, pp. 352-3.

30. STEPHEN L. et al. Dictionary of National Biography (ref. 22), pp. 237-41.

31. OFFLER H.S. Medieval Historians (ref. 6), p. 13 and n. 30.

32. ARNOLD T. (ed.) Symeonis Opera (ref. 3), vol. 1, pp. 139-40.

33. BILSON J. Durham cathedral: the chronology of its vaults. Archaeological Journal, 1922, vol. 79, 109-14.

34. ARNOLD T. (ed.) Symeonis Opera (ref. 3), vol. 1, pp. 247-61.

35. BATTISCOMBE C.F. (trans.) The relics of St Cuthbert, pp. 99-107. Dean and Chapter of Durham, Oxford, 1956.

36. RAINE J. (ed.) Libellus de virtutibus (ref. 13), chs. 40-3.

37. BATTISCOMBE C.F. (trans.) Relics of St Cuthbert (ref. 35), pp. 107-14.

38. HAMILTON N.E.S.A. (ed.) Willelmi Malmesbiriensis monachi De gestis pontificum Anglorum libri quinque, pp. 275-6. London, Rolls Series, 1870.

39. ARNOLD T. (ed.) Symeonis Opera (ref. 3), vol. 1, p. 141.

40. ARNOLD T. (ed.) Symeonis Opera (ref. 3), vol. 1, pp. 140.

41. CARVER M.O.H. Early medieval Durham (ref. 8), pp. 15-16.

42. YOUNG A. William Cumin: border politics and the bishopric of Durham 1141-1144. Borthwick Papers no.54, University of York, York, 1979.

43. ARNOLD T. (ed.) Symeonis Opera (ref. 3), vol. 1, p. 167.

44. RAINE J. (ed.) Liber de virtutibus (ref. 13), pp. 154-7.

45. STEPHEN L. (ed.) Dictionary of National Biography (ref. 22), vol. 16, pp. 453-9.

46. SCAMMELL G.V. Hugh du Puiset, bishop of Durham. Cambridge University Press, Cambridge, 1956.

47. ARNOLD T. (ed.) Symeonis Opera (ref. 3), vol. 1, p. 168.

48. RAINE J. (ed.) Historiae Dunelmensis scriptores tres: Gaufridus de Coldingham, Robertus de Graystanes, et Willielmus de Chambre, p. 11. Surtees Society, Durham, 1839.

49. SNAPE M.G. Documentary evidence (ref. 2), p. 23.

50. RAINE J. (ed.) Scriptores tres (ref. 48), pp. 11-12.

51. KNOWLES D., HADCOCK R.N. Medieval religious houses: England and Wales, 2nd edition, p. 357. Longman, London, 1971.

52. BONNEY M. Lordship and urban community: Durham and its overlords, 1250-1540, pp. 28-9. Cambridge University Press, Cambridge, 1990.

The building of the cathedral

C. DEVONSHIRE, BSc (Eng), FCIOB, MBIM, Senior Lecturer, University of Northumbria at Newcastle, and R. WILKIE, MSc, MCIOB, Principal Lecturer, University of Northumbria at Newcastle

SYNOPSIS. It is interesting to look at a major historical building and to compare the way it was built with the way the same design building might be built today. Thus the authors sought to compare the original construction of Durham Cathedral with how the same building would be built now, i.e. starting 900 years after the actual start.

HISTORICAL INVESTIGATION

1. Although there is little information relating to the original building of the Cathedral, some reasonable ideas of the construction can be obtained indirectly. This lack of real evidence of the actual construction process is a result of:

(a) the Client, in this case the Bishop and his associates, having no interest whatsoever in the process of the building, but only in the end result, so that no records of building methods would have been kept,

(b) the craft guilds discouraged others from observing their work, or recording it, in order to preserve the mystique of their skills.

(c) the loss, by time and dampness, of such records as would have been kept at the time, probably only of a limited financial nature

2. In the case of this cathedral, it has been suggested that some muniment records were destroyed deliberately by a steward, for his own advantage, while later other records were lost due to dampness and neglect.

3. Thus it has been necessary to piece together from many sources some possible ideas of the original building process.

THE METHODS

4. The original building work was begun at a time when there had been an invasion, followed by the arrival of William the Norman in the region to punish the local people for a rebellion, with, it is said, the punitive murder of many men and boys from York to Durham, an act for which there are well recorded precedents! However, the work was done throughout in a context of very cheap labour; indeed some of the labour would have been free. It can be seen from Boldon Book that within 15 miles of Durham, a reasonable distance to expect people to walk to perform duty to their Lord, there existed a significant number of people with duty of service of two or three days per week to the Bishop. He would probably have put some of this free labour at the disposal of the builders of his cathedral.

5. The Stephenson translation of Simeon states that St Calais ordered the existing White Church to be demolished before his return from France, and there is also a general view that St Calais returned with drawings for his new church. There is some doubt about both of these theories because:

(a) there is evidence to suggest that the White Church stood in part of the area now occupied by the cloisters and the chapter house. It is unlikely that a Bishop would have this existing church demolished, thus depriving the people of their place of worship for a period of several years. There was no need to clear the cloister area until much later in the building campaign. It has been suggested by Mr Currie, Cathedral Architect, that such demolition was only partial.

(b) there is no evidence of any specific plans in Durham, or indeed elsewhere at that time. For example, the famous drawings by Villiard de Honnecourt are no more than artistic non-scalar impressions of desirable external views. An example of the practice during that period can be found in evidence of a contract to build a church at Caterick, North Yorkshire. The documentation, of a type and layout not dissimilar to those used today, makes no reference to any diagrams, drawings or sketches. A written description of the works with specific references to overall plan dimension and requirements for features such as French buttresses is all that was provided.

6. Inspection of the walls shows very few individual stones not capable of being manhandled into position. Some of the large lintels would have needed mechanical assistance (block and tackle) to enable positioning but the maximum weight of any individual stone would be under 0.5 ton.

7. One can only speculate as to the likely methods used for moving the stone from the quarry situated to the South on the opposite bank of the river Wear. Oxen or horse carts could have been used. Illustrations of these methods show stone loads which can be estimated to be approximately 1.5 tons. Over level surfaces this would be feasible, but up inclines the load size could have been reduced. To build this cathedral, stone had to be hauled up steep tracks cut into the hillside. The tracks used might well be those existing today as footpaths, but are shown as vehicular tracks in illustrations circa 1700 AD.

8. Similar estimates can be made for other operations in the original building, based upon evidence from different sources. It is thus possible to make an estimated programme for the original works which is comparable to the programme for the works if built now. In the same way, it is possible to identify the likely methods of scaffolding, hoisting, etc., and this has been used for the general comparison given below.

Fig. 1. Comparison With Current Practice

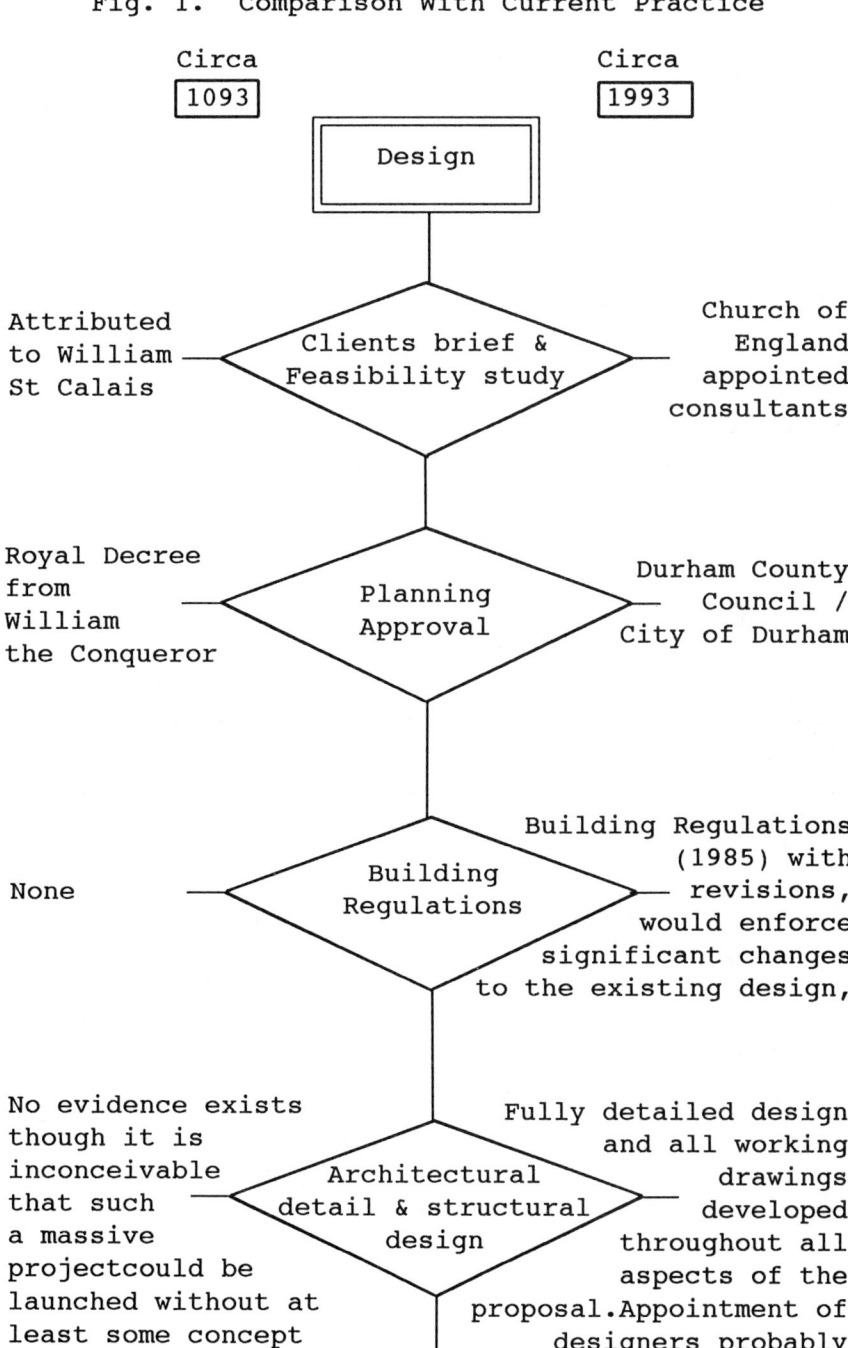

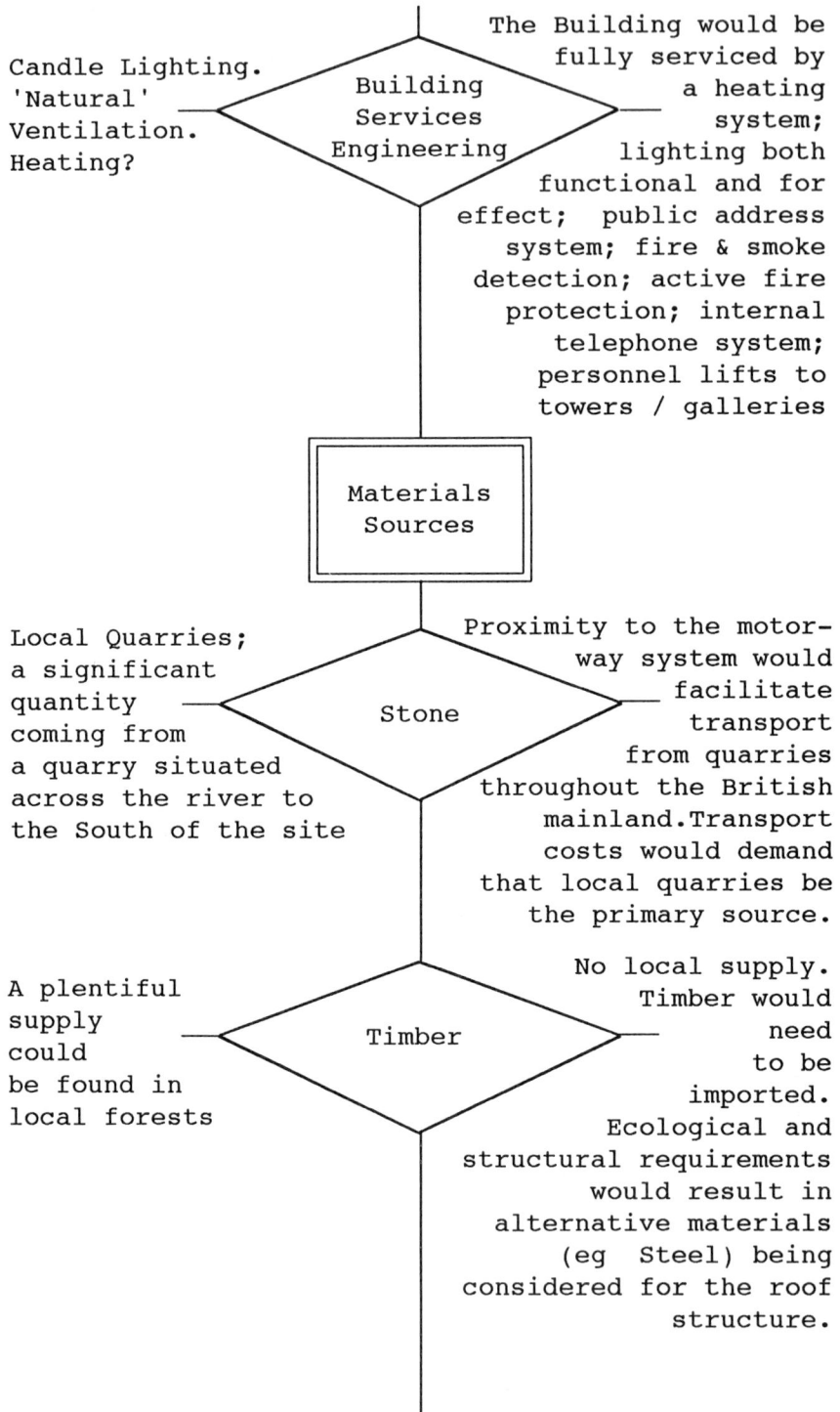

BUILDING THE CATHEDRAL

Successive master masons directing both design and practical site work	**Management and Labour**	A range of alternatives exist. Probably a project management approach being used to control both design information and contractors progress. (ref. Fig. 2.)
Reliance on well established / traditional craft skills and practices	**Quality Assurance**	BS.5750
Integration of imported labour into local population. Housing needs (labour camp) perceived as a development of the provision for the then expanding local community	**Site Set Up**	Inadequate number of local craftsmen would require supplementing by imported labour despite daily travel from the urban conurbations of Tyneside and Teesside and the whole of County Durham being practical
- a compound area for the storage of materials and stone dressing		- the establishment of site offices and materials compound on the palace green
- stone dressing sheds often built as 'lean to' against the previously completed walls of the cathedral		- all canteen, toilets and health and welfare provision to be in accordance with relevant regulations

21

	Labour Force	
No evidence of the likely number of people employed exists. Estimates vary, but a peak labour force of 300 to 400 on site appears likely. This number would be supplemented by quarry workers and material suppliers		An estimated 220 at peak, (average 148)

	Skills Provision	
Training was provided on site, with craft skills often being passed from generation to generation		Less reliance on site skills with mechanised stone cutting and dressing taking place at the quarry or other 'factory' locations

	Contract Period	
149 years 1093 - 1242		Estimated at 7 years 1993 - 2000

Methods

	Foundations	
Excavated by hand Disposal of the relatively small quantities of soil would probably have been by tipping down the banksides leading to the river Wear		Machine dig and cart away by truck to tip outside the city - Reinforced concrete foundations cut into rock base.

Left (traditional)		Right (modern)
– Mortar bonded hand built stone footings off rock shelf		Concrete craned or pumped where not possible to chute direct from pre-mix delivery truck. Reinforcement prefabricated at ground level and craned into position
Stone blocks individually hoisted and manhandled into bedded position. Fine dressed in situ. Mortar mixed on site.	**Walls**	Fully pre-dressed stone blocks hoisted on pallets and mechanically handled to position. Mortar by 'Megamix' or similar site set up
Timber centering man hoisted to position (see hoisting)	**buttresses and arches**	Fabricated alloy centering craned into position
Timber structure assembled in situ after man hoisting (see hoisting)	**Roofing**	Prefabricated sections craned into position
– slating? stone slab?		– slating techniques have changed little during the intervening centuries

Access & Working platforms

Lashed poles supporting a wattle hurdle platform to a height equal to the length of available poles (approx 9m). Above this level the timber scaffold would be cantilevered out from the wall under construction. Walls would also be built without the use of any tempoary working platform; the masons standing on the inner rubble fill of the lower wall section.

Alloy tubular independent scaffold erected to both sides of walls and tied through openings to provide fully secured access. This scaffolding (or proprietory system) would be <u>designed</u> to carry the planned loads

- gantry access for tracked powered trolleys for distribution from hoist to laying position.

Hoisting & Lifting

Block and tackle for lifting mortar skips and rope net slings for masonry.

- Capstans mounted at upper levels for lifting heavier loads.

- Manpowered treadmills mounted at upper levels

- hods and ladders

Track mounted saddle jib tower cranes. (Lorry mounted telescopic jib crane to North West and South East corners)

- platform hoists for masonry

- personnel hoists

Prefabrication

None (excluding centering which would be hoisted using a capstan or treadmill).

Little potential exists other than with the roof structure. <u>All</u> work that could possibly be done 'off site' would be.

BUILDING THE CATHEDRAL

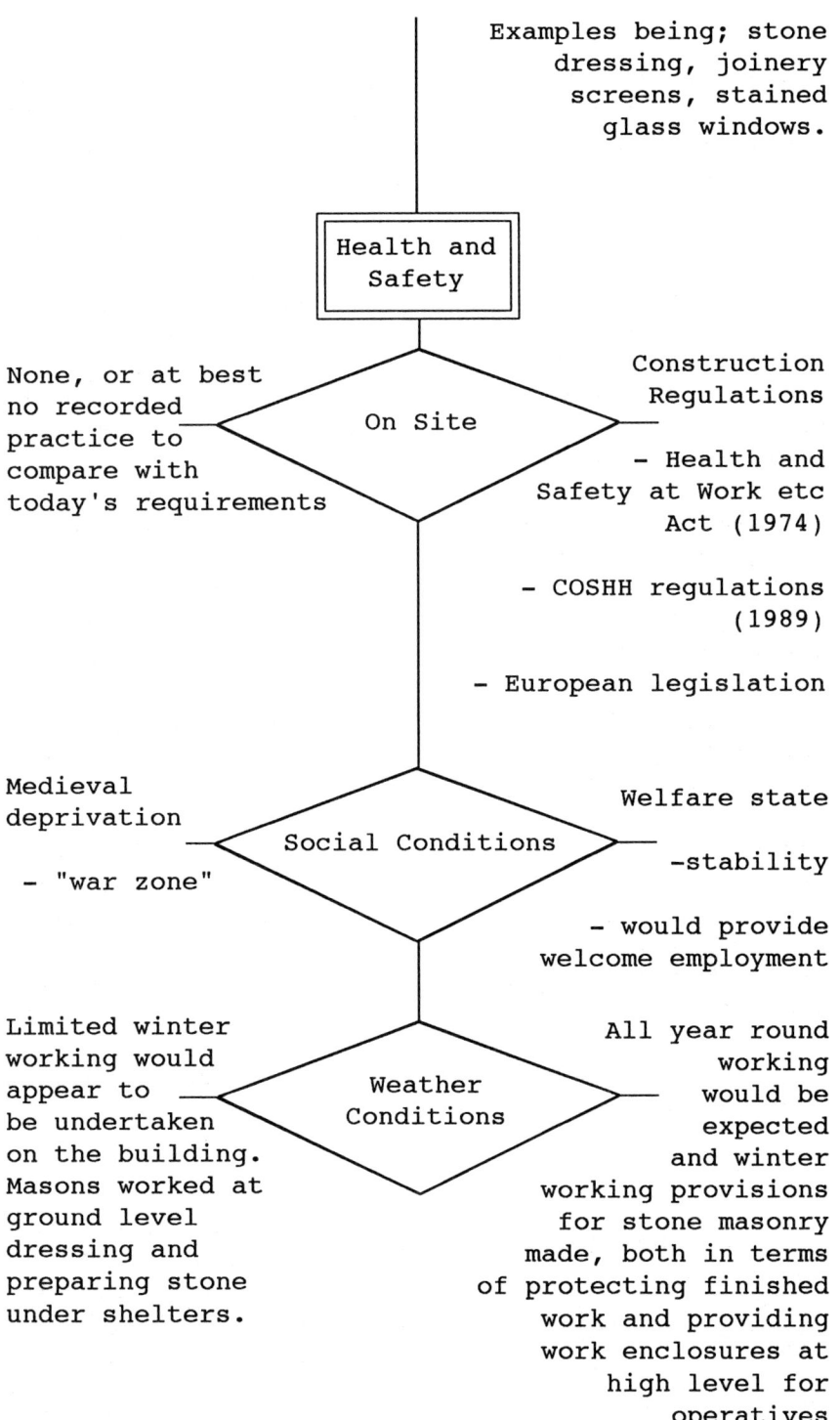

Examples being; stone dressing, joinery screens, stained glass windows.

Health and Safety

On Site
- None, or at best no recorded practice to compare with today's requirements
- Construction Regulations
- Health and Safety at Work etc Act (1974)
- COSHH regulations (1989)
- European legislation

Social Conditions
- Medieval deprivation
- "war zone"
- Welfare state
- stability
- would provide welcome employment

Weather Conditions
- Limited winter working would appear to be undertaken on the building. Masons worked at ground level dressing and preparing stone under shelters.
- All year round working would be expected and winter working provisions for stone masonry made, both in terms of protecting finished work and providing work enclosures at high level for operatives

25

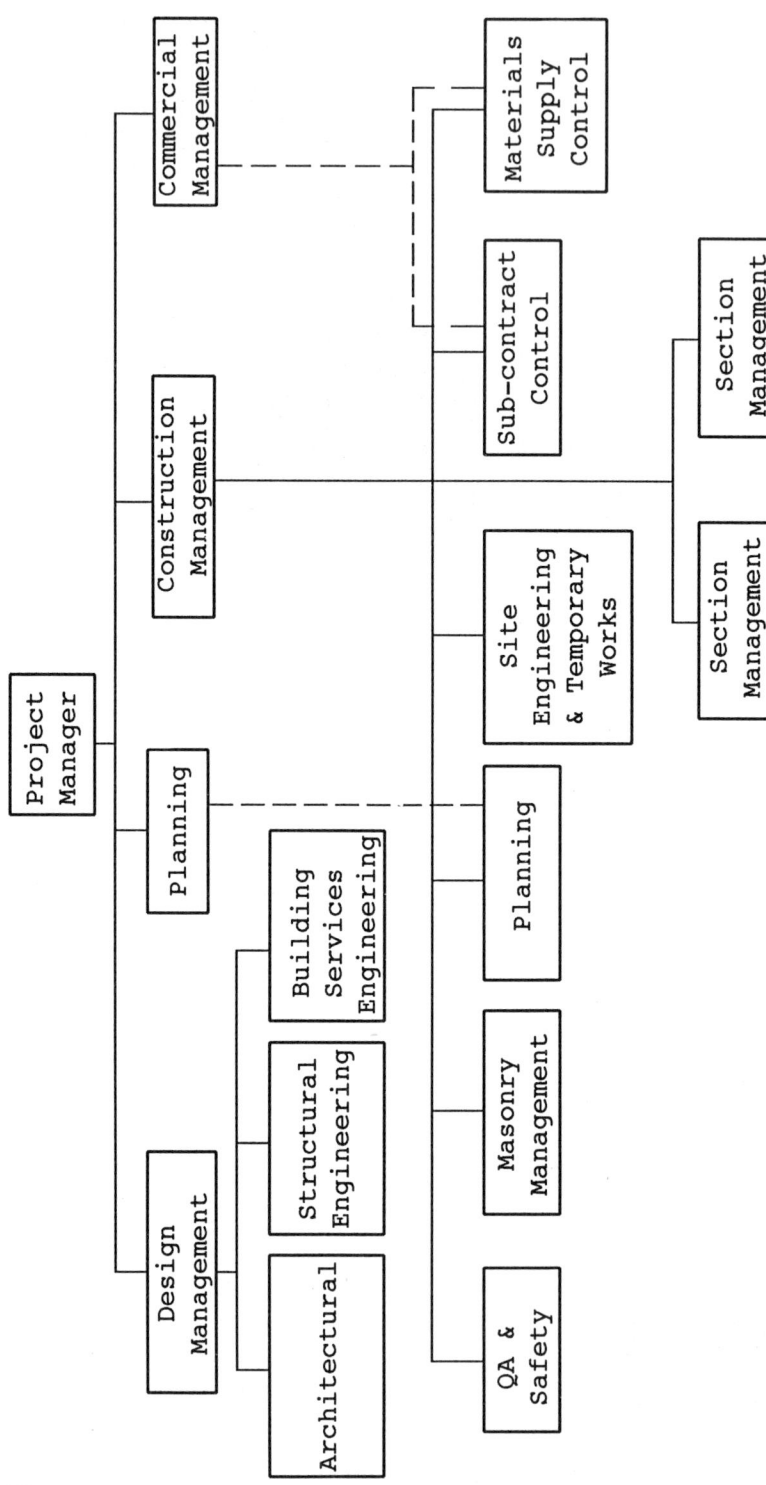

Fig. 2. Proposed Organisation Structure

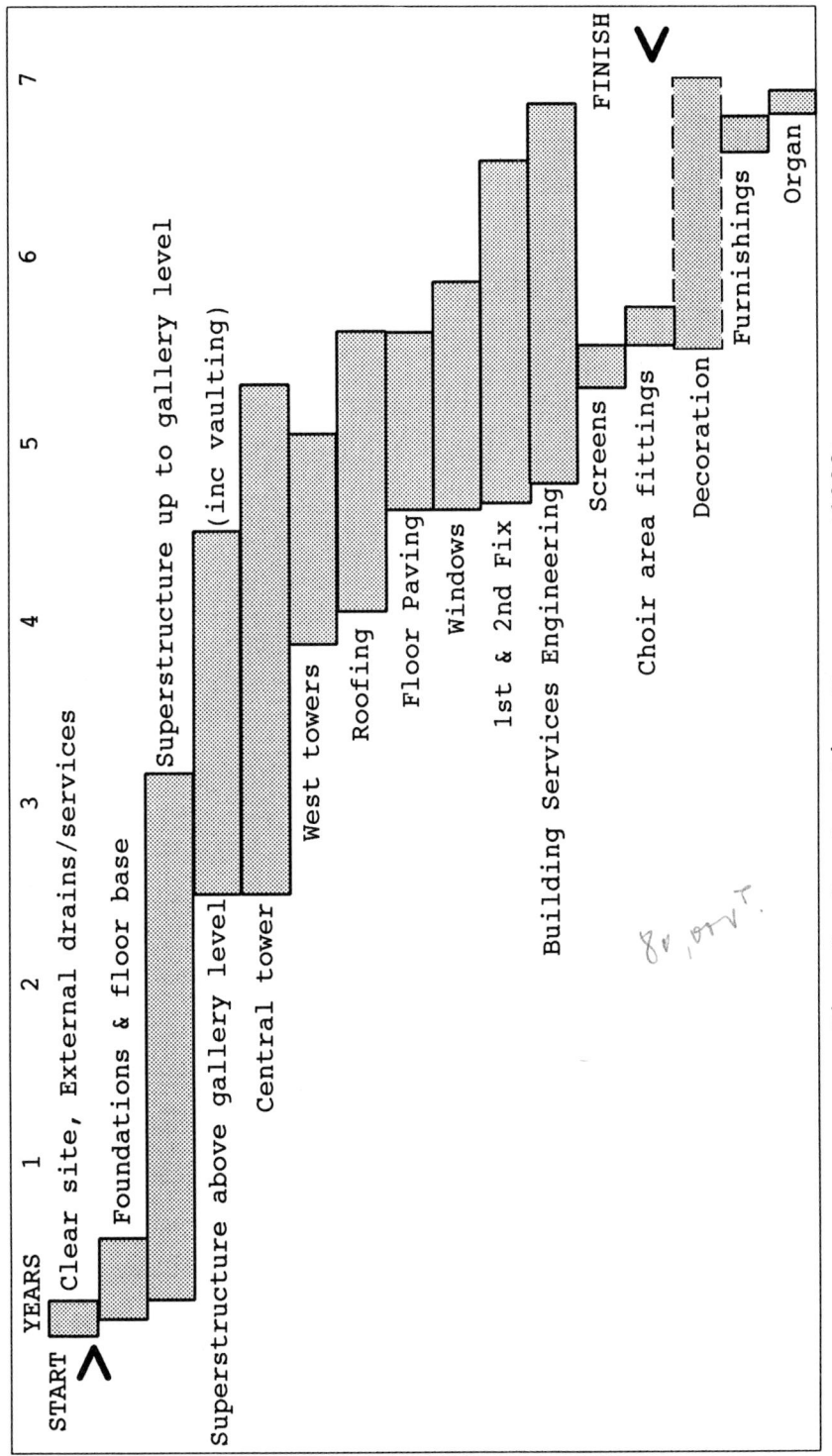

Fig. 3. Construction Programme (1993)

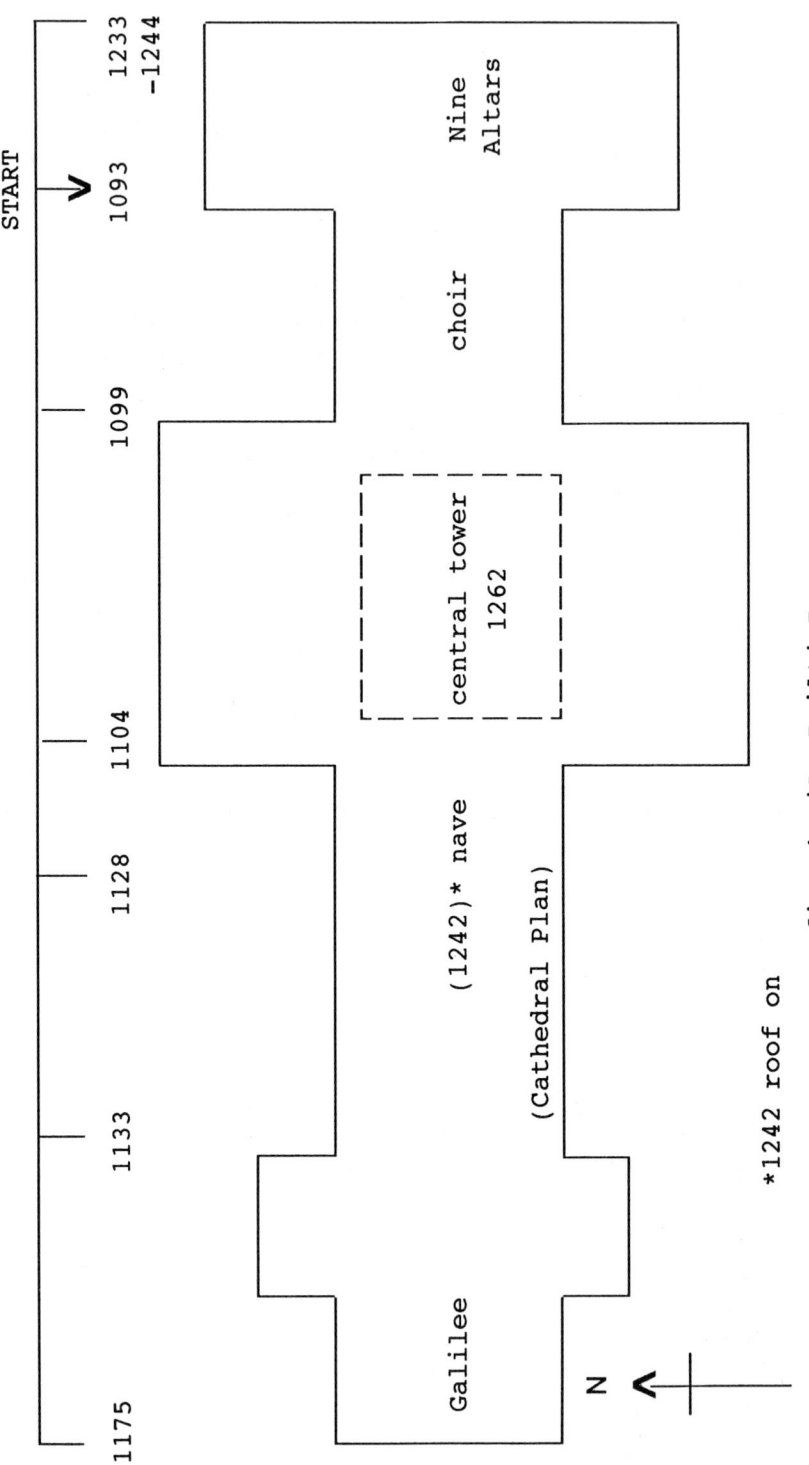

fig. 4. 'As Built' Programme

CONCLUSIONS

9. Many Fine Buildings have come down to us from our ancestors. They have stood the test of time, albeit with their occasional little difficulties, such as the foundation failure of the first attempt at a Lady Chapel on the North East side of this cathedral, or the occasional strike by lightening, at least twice on the tower at Durham. It is also true that our ancestors' failures have not only fallen down but, in most cases had the grace to disappear, under later buildings or whatever.

10. However this comparison of the 149 years it took to build the original main building, (i.e. not including all the extensive additional buildings and facilities we see today), with the 7 years in which those same buildings would be erected now, shows clearly not only the advance made in the building process over that time, but also the enormous rise in client expectation over those centuries. People now want their buildings more quickly, and in addition more cheaply in real terms, and they expect them to incorporate a higher standard of equipment, amenity and services than before. It is right that this should be so. Indeed it is the expertise of the engineer and the builder that makes it possible.

REFERENCES

1. AGRICOLA   De Re Metallica, trans HC and LH Hoover.   Dover, New York.
2. ANDREWS F.B.   The Mediaeval Builder and His Methods, Trans Birmingham Archeological Soc, 1975.
3. Bayeux Tapestry, some illustrations.
4. BEDE   The History of the English Church and People, trans Sherley-Price L,   Penguin.
5. BENNETT H.S.   Life on the English Manor 1150 - 1400, Sutton 1987.
6. BILSON   J.   Durham Cathedral, the Chronology of its Vaults, A.J. 1922.
7. Bolson Book,   trans Morris, 1982.
8. CARTER J.   Some Account of the Cathedral Church of Durham, London 1801.
9. COLVIN H.M.   The History of the Kings Works, Vol 1 & 2, HMSO 1963

10. CURRY I. Sense and Sensitivity, Durham Cathedral Lecture, 1985.
11. GERVAISE. His History of the Burning and Repair of the Church of Canterbury, circa 1174, trans Rev Robert Willis, reprinted in Architectural History of Some English Cathedrals, 1972.
12. GILES F. & J. Life in a Mediaeval Village, 1989
13. HARVEY JOHN, Mediaeval Craftsmen, Batsford, 1975
14. HARVEY JOHN, Mediaeval Builder.
15. HAYMAN J. How to Design a Cathedral, Some Fragments of the History of Structural Engineering, Proceedings of Institution of Civil Engineers, Feb 92
16. INNOCENT C.F. The Development of English Building Construction, Cambridge 1916.
17. PEVSNER N. 2nd Edition.
18. POOLE A.L. Obligations of Society in 12th & 13th Centuries, The Ford Lectures at Oxford, 1944.
19. PROUD K. The Prince Bishops of Durham, 995 to 1539.
20. Rites of Durham, pre 1600, Surtees Society, 1903.
21. RODWELL W. Church Archeology, Batsford / English Heritage, 1989.
22. ROLLASON D.W. (Ed), Cuthbert, Saint & Patron, Dean & Chapter of Durham 1987.
23. SALZMAN L.F. Building in England Down to 1540 Oxford UP, written 1934, first published 1952.
24. STRANKS C.J. The Venerable, Pictorial History of Durham Cathedral, 1965.
25. SYMEON OF DURHAM, A History of the Kings of England, trans Joseph Stevenson, 1858.
26. TAYLOR A.J. The King's Works in Wales, 1277 - 1330, HMSO, 1974.
27. TAYLOR H.M. & TAYLOR J. Anglo-Saxon Architecture, Vol 1 & 2, Cambridge University 1965.
28. VITRIUS The Ten Books on Architecture, trans Morgan M.H. Dover, New York.
29. English Historical Documents on the trial of William St Calais in the Court of William II, Circa 1088

ACKNOWLEDGEMENTS

Dr & Mrs Rollason, whose encouragement made it all bearable.

Mr R.C. Norris, Deputy Chapter Librarian.

Ms J. Plinston, University of Northumbria at Newcastle.

An appraisal of the topology and loading of a section of the vaulted roof of Durham Cathedral

D. M. LILLEY, BSc, PhD, CEng, MICE, Lecturer in Structural Engineering, Department of Civil Engineering, University of Newcastle-upon-Tyne

SYNOPSIS. A study has been made of the shape of a section of the vaulted roof of the Cathedral immediately to the West of the Main Tower. The investigation was undertaken using photogrammetric surveying techniques; two photographs were taken from the floor of the Nave and the precise positions of 384 points on the vaulted roof were obtained. Computer software was used to formulate the shape of the roof from the survey information. Further analysis suggests that the section of vaulted masonry (and associated fill material) exerts a vertical load of about 212 tonnes on the supporting walls and columns.

INTRODUCTION
1. The initial aim of this study of the vaulted roof of the Cathedral was to gain a detailed understanding of the structural morphology and loading regime within one section of the roof. This information could then be developed into a sophisticated analysis of the masonry structure from which its behaviour could be examined under the existing load conditions. Interest was originally focused on the likely values of stress and deformation within the vaults and how these may have changed with time after construction.
2. At the present time, some of these objectives are still to be realised; work is continuing on the structural analysis of the vaults.
3. It is very easy for modern "intelligent" man to be dismissive of the ways and thinking of previous generations. A close examination of the shape of the vaulted roof, however, reveals a complex shape constructed to a level of accuracy which would be readily acceptable within modern working practice.
4. About 900 years ago, the people in the Durham area had very simple (and often short) existences in a world in which life itself could be curtailed very quickly by disease, acts of violence or other misfortune. Very few people could read or write and probably the main objective of the inhabitants of this area was simply that of survival.

TOPOLOGY AND LOADING OF ROOF

5. In the light of this way of life, Durham Cathedral was built like other similar religious buildings using a combination of artisan and unskilled labour. Existing literature (ref. 1) suggests that the main body of the Cathedral was erected over a period of about 40 years. It is doubtful (but still possible) that any one individual would have seen the construction of the main building from the laying of the foundation stone in August 1093 to its completion.

6. To call the people who built the Cathedral "engineers" is probably a misuse of the term. As far as is known these people had no sophisticated measuring devices, and certainly no "accurate" theodolites, levels or any of the gadgetry which modern engineers/surveyors have at their disposal. In contrast, the ancient builders must have been able to communicate very well (using the spoken word) between themselves, their contemporaries and the generations which immediately preceded or followed their own.

7. Two major influences were probably present in the minds of those working on the construction of the Cathedral. The first was that religious belief at that time would have insisted that God's House was built to the highest standards which were then possible. The second influence is that the Normans, having substantially defeated the Saxons, would have wanted to consolidate their presence and demonstrate their construction abilities and "management techniques" to the indigenous population, confirming their power over the land and its people.

8. Whichever thoughts prevailed, there seems little doubt that all efforts would have been made to ensure that the construction of the Cathedral was as perfect as possible. In the early days of the construction there must surely have been work completed but subsequently rejected as being less than "perfect". Any such problem with masonry work for the vaulted roof would have been resolved on the ground, as a "trial" erection at low level would have been made before final construction was attempted.

9. Not surprisingly, no information is available about checks made on the accuracy of the construction, a possible method of which is described in ref. 2. It may have been that such was the relief that the walls and roof remained in place after the supports were removed that minor dimensional tolerances were ignored. Over a period of time any differences between the existing and the "planned" structures may have changed; there is no way of retrieving information about the original profile of the vaulted roof.

THE SHAPE OF THE VAULTED ROOF

10. A major problem to be resolved before any attempt can be made to analyze the existing loads within the vaults is that of determining the topology of the masonry structure.

11. Access to the lower surface (intrados) of the vaults

is not possible without scaffolding, but this would prove expensive and impractical given the everyday life and use of the Cathedral. Inspection of the upper surface of the vaulted roof revealed little information; in the roof to the West of the Tower the shape of the vaults is obscured by a considerable depth of fill material. Similar material was removed from the vaults of the East side a few years ago, but the task of determining the profile of the vault by "conventional" surveying would still have been difficult owing to the three-dimensional curved shape of the structure.

METHOD OF SURVEY

12. Several advanced methods of surveying were contemplated, including that of electronic distance measurement. A major concern was the need to cause as little disturbance to the normal proceedings of the Cathedral as possible. This implied that time spent taking measurements should be reduced to a minimum but satisfying the conflicting interests of both survey personnel and Cathedral authorities.

13. After a preliminary inspection of the vaulted roof of the Cathedral from the floor of the Nave and then from the galleries within the upper walls, it was evident that the most practical method of obtaining details of the profile of the roof was to use the method of photogrammetry, commonly used for aerial surveying.

14. This technique has been practised and well-documented (ref. 3) for many years. Its main advantage in respect to its application within the Cathedral is that there is no need for personnel to be in close proximity to the vaulted roof itself; all measurements can be made from photographs taken with a high performance camera in a relatively short period of time from the floor of the Nave.

15. The principle of photogrammetry requires two photographs to be taken of the same object from different viewing positions. Detailed information is required about the exact location, height and attitude of the camera when each photograph is taken. The photographs themselves are taken using very high quality lenses and black and white film, and are developed and printed to high photographic standards.

16. Two such photographs were taken from the floor of the Nave in the area under the vaulted roof immediately to the West of the Tower and are presented as Figs. 1 and 2. When viewed separately, the casual observer will see little difference between the two photographs except that the viewing position is different in each case. However, when placed in a stereoscope, the two photographs can be merged to become a stereoscopic pair, producing an image which

TOPOLOGY AND LOADING OF ROOF

Fig. 1. View of section of Nave vaults from the first position.

Fig. 2. View of section of Nave vaults from the second position.

appears to be three-dimensional. The process by which this is achieved is a reproduction of that of the human brain in its combination of visual images from each of the two eyes separated by a small distance (about 65 mm) on a human face, but in the photographic case the separation is much greater.

17. Modern equipment allows the operator of the stereoscope to position a moveable marker (cursor) which is visible within a three-dimensional image of the object. Adjustments made within the instrument allow the apparent vertical position of the cursor to be changed; the aim of the operator is to adjust the instrument and place the cursor so that it appears to be "resting" on the object being viewed, ie. the cursor and the object appear to the operator to be at the same height. The apparent height and horizontal position (measured in plan) of the cursor are then recorded within the system when the operator is satisfied that the height of the object matches that of the cursor.

18. The skill of the operator is paramount at this stage if accurate results are to be obtained. Some individuals have a greater ability than others in comprehending the stereoscopic effect created by this technology. The reasons for this difference in personal ability are not immediately obvious and are not for discussion here, but acknowledgement is due to the staff of the Department of Surveying at the University of Newcastle-upon-Tyne for their invaluable assistance with this part of the study.

19. Vertical heights at 374 separate points on the underside of the vaulted roof were determined using the equipment and expertise available at Newcastle. This number of points arose after it was decided to obtain values of profile height at positions based on a horizontal grid of 0.5 metre spacing. In reality the number of points represents a reasonable compromise between the conflicting needs for accuracy and economy. It then became possible to construct a database of three-dimensional Cartesian coordinates representing the lower surface of the vaulted roof.

COMPUTER-GENERATED ROOF PROFILES

20. The next stage of the study investigated the profile of the vault in greater detail using the information obtained from the photogrammetric study in conjunction with SURFER (ref. 4), a sophisticated computer software package.

21. This software enables graphic visualisation of three-dimensional surfaces and structures from any given viewing position and also offers the opportunity to provide contours of vertical height based an enhanced database. Fig. 3 illustrates the profile of the first section of the vaulted roof to the West of the Tower viewed from the South-West and at angle of 30° above the horizontal.

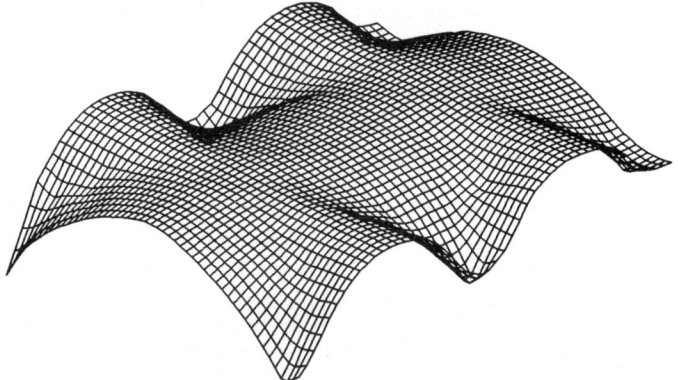

Fig. 3 Computer-generated profile of first section of vaulted roof

22. The computer software allows the viewing position to be changed so that the different geometries of the arches within the vaulting can be considered. Several views of the structure are presented in Figs 4-7.

Fig. 4 Vaulted roof viewed from the East

Fig. 5 Vaulted roof viewed from the North

Fig. 6 Vaulted roof viewed from the North-East

Fig. 7 Vaulted roof viewed from the South-East

23. Figure 8 illustrates the contour heights of the surface of the vaulted roof, and was obtained from the grid determined by the SURFER program. This picture relates the same essential information about lines of equal height as an Ordnance Survey map, except in this case the contours represent differences in vertical height of 25 mm. In essence, the closer together the contour lines, the more vertical is the surface of the vault. Hence the darker areas in Fig. 8 represent the regions where the level of vault is dropping to meet the supporting columns. The contours for the areas directly above the columns have been removed from Fig. 8 for reasons of clarity; some distortions appear in the mapping process if an attempt is made to draw too many contour lines within a short horizontal distance.

24. It is apparent from the contours that, even after the passage of nine centuries, the shape of the vault is very nearly symmetrical. Small deviations in the "flat" area of the vault can be seen but the indications are that the deviations in vertical height are no more than about 25 mm from the intended shape. It is possible that the rendering on the intrados has covered the imperfections of the vaulted masonry, but the inverse may also be true since it may be that the rendering has been applied to a greater thickness in one place than at another.

DEPTH OF CONSTRUCTION

25. A small number of openings have been made in the vaulted roof to allow ropes, etc. to be connected to the chandeliers above the Chancel. Other larger openings have been made at some time in the past and may even have been in

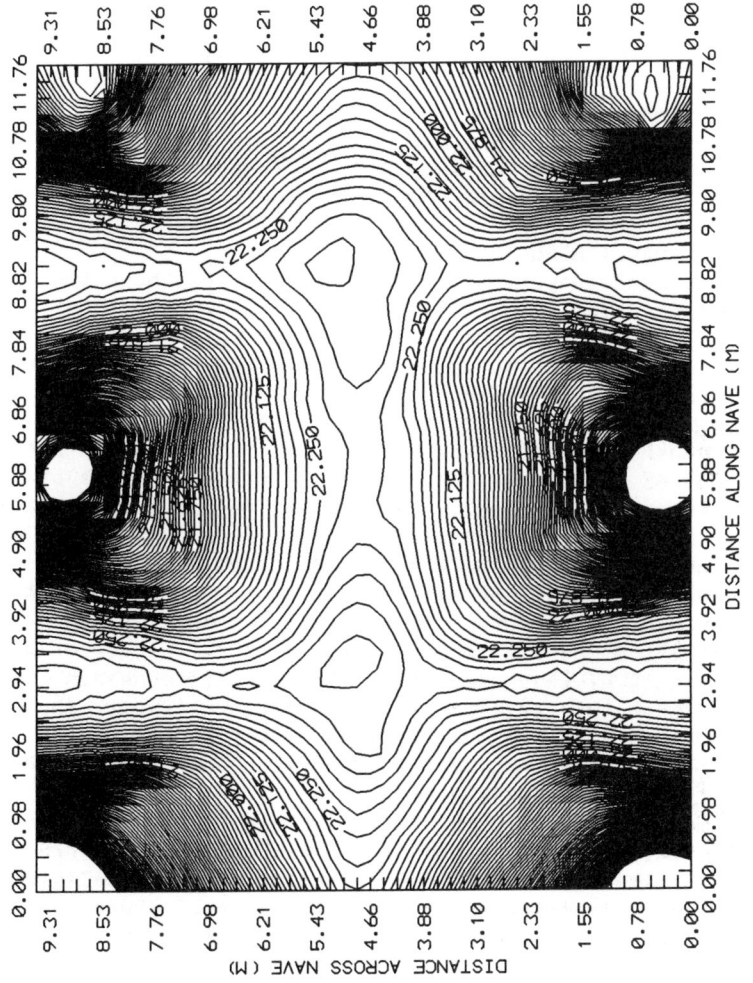

Fig. 8 Vertical contours of vaulted roof - contours at 25 mm intervals

the original scheme. There is one at the East end of the Cathedral which is presently used to allow powerful electric lights to illuminate an area of the Cathedral underneath. There is another at the West end although this appears to be unused at present. Both of these openings are about 1 metre in diameter and are formed in the centre of the vaults so as not to interrupt their visual appearance.

26. The depth of the vault structure can be estimated from an inspection of these openings. Allowing for the decorative masonry on the underneath of the vault, the depth of masonry which forms the vault appears to be between 300 and 400 mm. This dimension would have been important to the original masons as it has implications regarding the ease of handling individual blocks of stone.

LOADING ON VAULT STRUCTURES

27. A visual inspection of the roof trusses above the vaulted roof revealed that in the roof space to the West of the Tower the bottom members of the main trusses spanning between the walls are not directly connected to the vaulted roof. The situation in the other roof void (East of the Main Tower) is slightly different as there are secondary timber members connecting between the top of the vault and the bottom of the truss. The reasons for the presence of these members is not clear; they are not required by either the roof truss or the vaulted arch.

28. The main loading on the vaults arises from their self-weight and the loose material which is now in position above the vault structure. In recent years the vaults to the East of the Tower have been cleared of this loose material, much of which was reported to be slate-debris discarded when this section of the roof was renewed. The voids created by the removal of this material are about 2 metres deep. To the West of the Tower the vaults above the Nave still contain debris of a similar nature.

29. Further study of the curved surface of the vault and the implementation of some of the utilities of the SURFER computer package allowed its surface area to be calculated. The surface area of the section of vaulted roof shown in Fig. 1 is 129.1 m^2. There is no structural significance in this value, although it may be of interest to a quantity surveyor if any form of surface treatment was being contemplated.

30. Of far greater interest to the structural engineer is information about the volume of material contained within one or more of the vaults. This volume has been calculated assuming that the surface of the fill material above the vault is approximately horizontal and level with the top of construction at the crown of the vault.

31. Three different methods of calculating the volume were used, each of which is a form of numerical integration.

The methods were those of the Trapezoidal Rule, Simpson's Rule and Simpson's 3/8 Rule and all gave similar values to within an accuracy of 0.01% of the computed volume.

32. The total volume of material above the vault (including the structure itself) was calculated to be 94.1 m^3, assuming a depth of construction of 400 mm. If this assumed depth is increased to 600 mm, then the volume of material becomes 116.9 m^3. Further, if it is assumed that the density of the sandstone masonry is 2300 kg/m^3 and a similar value is assigned to the fill material, the entire weight of the material in this section of the roof (using the smaller construction depth) is approximately 2123 kN or approximately 212 tonnes. If the larger depth is assumed, the weight of material becomes 264 tonnes. These figures are likely to over-estimate the true values; not all the areas of the vault are exactly full to the same level and the density of the fill material, although compacted naturally over the years, is probably less than that of sandstone.

33. All of the vertical dead load represented by this mass of material has to be transmitted ultimately to the foundation via the upper walls of the Nave, and the massive columns which line that area of the Cathedral.

34. The walls in the upper section of the Nave have arched openings within them, and this fact together with the architectural arrangement of the ribs suggests that the load is transmitted to the springing points of the ribs and vertically down through the columns, thus avoiding the arched openings.

35. The weight of material in and above the vaults may seem considerable to the casual observer, but in engineering terms, the compressive stresses induced in the massive columns and masonry directly above these columns are small relative to those which would cause failure of the masonry. The horizontal forces produced by the vaulted structure and which are resisted by buttresses and other structural components are of greater structural significance and are to be examined in a later stage of this investigation.

PROPOSED FINITE ELEMENT ANALYSIS

36. Although guidance on conventional analysis of masonry-arched structures can be found from existing literature (eg. ref. 5), it was felt more appropriate to use modern computer techniques to investigate the distribution of stress and values of deflection within a section of the vaulted roof. Consequently the next stage in the study is to develop a mathematical model of the roof using additional computer software incorporating finite element analysis. This work is forming part of a current research project supervised by the author, the results of which will be published later.

CLOSING COMMENTS

37. The visually-impressive construction of the vaulted roof of the Cathedral has been examined by a modern surveying technique and found to have few deviations from its likely original shape. This fact is made more remarkable by the age of the structure and also that the apparent maximum deviation in shape is approximately 25 mm. This accuracy of construction is similar to that to be expected using modern methods of construction and associated technology.

38. General observation from the Nave suggests that some apparent distortion has occurred in the alignment of the arch ribs which pass diagonally across the vaulted roof. This movement appears to have been mostly in a transverse horizontal direction, and it may give an indication of the sequence of construction of the vaults. The distortion appears to be most pronounced in the end section of the vault roof, ie. that which has been studied above. It is possible that this was the first vaulted section to be built and that adequate provision was not made for the horizontal forces created by its construction. As a consequence, the adjacent arch rib became displaced horizontally but did not collapse. This fault, if it had been noticed, would have been difficult to correct without demolition and rebuilding. Assuming this to be the case, it is likely that construction of the remaining vaults took place following the provision of strengthened horizontal restraints.

ACKNOWLEDGEMENTS

39. The work and expertise of Mr. Hugh Buchanan and Mr. Ian Newton of the Department of Surveying at the University of Newcastle-upon-Tyne is gratefully acknowledged in the measurements of surface profiles of the vaults. The assistance of the Cathedral staff in obtaining guided access to parts of the building not normally accessible to members of the public is also gratefully acknowledged.

REFERENCES

1. STRANKS, C.J. This Sumptuous Church, SPCK, chap. 1, 1973.

2. PROWSE, W.A. Vision and Practice: Building the High Vault, Durham Cathedral Lecture, 1977.

3. BANNISTER, A. and RAYMOND, S., Surveying, Sir Isaac Pitman and Sons Ltd., London, chap. 11, 1972.

4. SURFER Version 4 Reference Manual, Golden Software Inc, Golden, Colorado, 1990.

5. HEYMAN, J. The Masonry Arch, Ellis Horwood Ltd., Chichester, 1982

The Romanesque high vaults of Durham Cathedral

M. THURLBY, PhD, FSA, Associate Professor, Department of Visual Arts, York University, Ontario

SYNOPSIS
1. In the Romanesque Cathedral at Durham high vaults were planned for the choir and transepts from the first. This plan was subsequently modified in the south transept. The nave was originally intended to be wood roofed. The diaphragm arches in the choir and transept galleries, and the quadrant arches in the nave galleries, were not conceived as abutments for the high vaults.

THE CHOIR HIGH VAULT
2. The Romanesque high vault of the eastern arm of Durham Cathedral was replaced in the 13th century as part of the building programme that added the Chapel of the Nine Altars after 1242. The vaults over the shrine of St Cuthbert were threatening ruin because of cracks and fissures (ref. 1). Strictly speaking this applies to the apse semi-dome and the forebay. It is therefore possible that the replacement of the entire choir high vault was an aesthetic decision to blend with the new work in the Nine Altars. The rood screen would have blocked the lower part of the choir elevation from view from the nave and therefore the initial impression upon looking east from the nave would have been of a new eastern arm.
3. The evidence for the Romanesque high vault, which was first presented by Hills, has been extensively discussed by Bilson (ref. 2). The case for high vaults is unequivocal given the traces of the vault trajectory from the capitals above the major piers to the arcs congruent with the clerestory window heads. This evidence is now supplemented with that above the minor piers which was examined in 1991 when scaffolding was erected for cleaning and re-limewashing the high vault. Lines of the former Romanesque vault were revealed and some fragments of webbing remain behind the Frosterley marble shafts that carry the 13th-century wall arches. Therefore

Fig. 1. Durham Cathedral, north transept, east elevation, from south nave gallery

the Romanesque high vault was definitely constructed. But what form did it take?

4. Bilson reconstructed a quadripartite vault over single bays with transverse arches above the major piers but not above the minor piers. This is surprising because there are three shafts on the gallery sill above the minor piers which would logically articulate a transverse rib and two diagonals. Nevertheless Bilson opted for a reconstruction based directly on the extant Romanesque high vault in the north transept where the there is no transverse rib above the minor pier, and the two diagonal ribs there are 'logically' articulated by two shafts (Fig. 1). Bilson's reconstruction has received widespread acceptance and his plan of the Romanesque cathedral is the one most often reproduced (ref. 3). It was challenged by John James who suggests that a sexpartite vault was erected over double bays (ref. 4). He appreciates that this does not accord logically with the three shafts over the minor piers, and he therefore attributes the sexpartite vault to a hypothetical change in plan. The evidence does not support James's reconstruction. What he perceived as asymmetry in the traces of the Romanesque vault is inevitable given the uniform articulation of the 13th-century vault on alternating Romanesque supports. Above the major piers the transverse rib is carried on the central shaft of the triplet that rises from the ground. To either side the next shaft takes the diagonal rib while the shaft that sits on the gallery sill supports the Frosterley marble shaft for the wall rib. Thus the three ribbed elements of the vault occupy just the three central Romanesque shafts of the five-shaft group, the three shafts that originally carried just the transverse arch of the Romanesque vault as in the north transept. The marble shafts that sit on the columns that carried the Romanesque diagonal ribs occupy only part of the space taken up by the original vault springer. Since the marble shafts rise straight up the wall the arc of the original vault is clear. Above the minor piers, the transverse rib of the 13th-century vault occupies the central shaft of the three-shaft group. The diagonal ribs are set on the outer shafts. There are no Romanesque shafts to carry the Frosterley shafts for the wall arches. Therefore they have to be corbelled out to the side of the vault springers. The three ribbed elements of the 13th-century vault here occupy all three Romanesque shafts rather than just the central three of five above the major piers. Hence the asymmetry in the evidence for the Romanesque traces which seems to be

Fig. 2. Noyon Cathedral, nave vault   *Fig. 3. Durham Cathedral, north transept vault*

exacerbated by the introduction of the corbelled marble shafts. These shafts mask the evidence for the trajectory of the Romanesque vault when viewed from the ground or from the gallery, but examination from scaffolding revealed traces of the original vault that gave a trajectory that mirrored that from the major piers. The clearest evidence is in bay S4 where fragments of the original webbing are extant. To apply this evidence to James's reconstruction would result in a sexpartite vault with intermediate 'ribs' as wide as the massive transverse arches above the piers. Clearly this is ludicrous. The obvious alternative is the one first proposed by Quekett and Cheetham, which was followed by Pevsner and Heliot, but only explained by Curry (ref. 5). Thus the triple shafts above the minor piers would carry the diagonals and a transverse rib. This would result in four-part vaults over single bays with alternating major and minor transversals in accordance with the major and minor piers such as that employed later in the nave of Noyon Cathedral (Fig. 2) (ref. 6).

5. We may be sure that the apse was rib-vaulted to complete the three-part division given by the wall shafts (ref. 7). For the forebay there is no direct evidence; there is a barrel in this position at Lastingham, but the use of ribs elsewhere in the cathedral may make a rib-vault iconographically more acceptable. Be that as it may, the cracks and

fissures in the vaults observed in 1235 may have resulted from the settlement of the Romanesque towers above the aisle apses which flanked the forebay (ref. 8).

6. The diaphragm arches across the choir gallery are often seen in connection with abutment for the high vault. This cannot be the case because the masonry in the spandrel of the diaphragm arch next to the gutteral wall stops level with the extrados of the arch (ref. 9). This is level with the abaci of the high vault capitals. There is no masonry above the diaphragm to abut the thrust behind the haunches of the high vault. In this regard the arches are precisely paralleled in the choir galleries at Chichester Cathedral, the main span of which was wood-roofed (ref. 10). The design is quite unsuitable as an abutment for the high vault. That it was not conceived in these terms is indicated by the responds of the diaphragm arches in the gallery at the chord of the main apse of Norwich Cathedral. Above the respond on the gutteral wall there remains the scar of the spandrel of the diaphragm arch and this continues to the gallery roof. This construction was used to provide abutment for the triumphal arch. Therefore had a buttressing job been intended for the Durham diaphragms the masonry would have continued above the spandrels next to the gutteral walls. As it is the diaphragms were designed in connection with the gallery roof. The only other possible structural purpose of the diaphragm arches is as a tie between the gutteral wall of the choir and the outer wall of the gallery. Given walls seven-feet thick at Durham lateral thrusts of the high vault are unlikely to have presented a problem (ref. 11).

THE TRANSEPT HIGH VAULTS

7. The transepts have been differently interpreted by Bilson and Bony (Figs. 3 & 4) (ref. 12). Bilson argued that they were initially intended for high vaults, though this plan was revised during construction. Preparations for the high vaults in both transepts were followed to the level of the east gallery arches. Construction of the south transept then continued in advance of the northern arm but with the abandonment of the plan for the high vault. The former vault shafts were carried up to the roof plate, and the design of the inner plane of the hollow-wall clerestory was revised to accommodate tall, minor arches flanking the large arch facing each window. This revised design was constructed but subsequently the present vault was

inserted. In contrast to the south transept, Bilson found no evidence in the north transept for the erection of a wood-roofed elevation prior to the construction of the present vault. However, he suggested that the design of this vault as a whole was not established at the outset; the support of the diagonal ribs on corbels on the west wall and the northern bay of vaulting spanning two bays of the east arcade seemingly represented revisions during construction (Fig. 3) (ref. 13).

8. Bilson's interpretation was challenged by Bony who contended that the asymmetry between the bay rhythm of the east and west walls of the transepts indicated that high vaults were not planned at the outset. In both arms the first two bays on the east are matched in width by their counterparts on the west wall, but bays 3 and 4 on the east stand opposite a single bay on the west. The reduction in the scale of the major piers from the choir to the transepts and the introduction of the clerestory wall passage in the transepts, as opposed to the solid wall at this level in the choir, were also cited by Bony as evidence that high vaults in the transepts were not intended from the first (ref. 14)

9. The articulation of the main arcade, gallery, and the aisles of both transepts is based on the choir. While the diameter of the minor piers is the

Fig. 4. Durham Cathedral, south transept vault

Fig. 5. Waltham Abbey, north nave elevation

Fig. 6. Durham Cathedral, south transept, west clerestory

same in the choir and transepts, the major piers in the transepts are smaller than those in the choir. This scaling down involved a reduction only in the length of the pier on the axis of the arcade; the thickness of the pier remains the same in both choir and transept. If the transept piers had been built the same size as those in the choir then the entrances to the chapels off bays 2E and 3E would have been reduced to nearly half their present width (Fig. 1) (ref. 15). The reduction in the size of the major piers in the transepts was therefore conditioned by the need to provide adequate width for the chapel entrances and not by the abandonment of a plan for the high vault.

10. The front plane of the gallery in both transepts is set back from the main arcade to accommodate paired shafts above the minor piers and single shafts flanking the major piers. The use of paired shafts above the minor piers in contrast to the triple shafts in this position in the choir reflects the reduction in length of the double bay from the choir to the transept. Otherwise the similarity of the design suggests that, like the choir, both transepts were intended for high vaults. Against this Bony argued that shafts starting at gallery level should not necessarily be equated with a vaulted design, and he cited wood-roofed nave of Waltham Abbey in support of his contention (Fig. 5). However, the analogy between the Waltham and Durham gallery-set wall shafts is flawed. At Waltham the

Fig. 7. Durham Cathedral, north nave elevation

Fig. 8. Durham Cathedral, north transept to NE

front plane of the gallery is not set back from the main arcade in the manner of Durham; the Waltham shafts do not sit on the gallery sill but rather are carried on head corbels. Furthermore, at Waltham there is just one shaft, as opposed to triplets and singles in the choir, and pairs and singles in the transepts, which logically articulate the vault ribs at Durham. The Durham scheme is more closely paralleled in three English Romanesque churches known to have had high vaults: the nave of Gloucester, the choir of Malmesbury, and most directly in the nave of Lindisfarne Priory where paired shafts on the gallery sill carried diagonal ribs of the high vault exactly as in the Durham north transept (ref. 16).

11. Bilson's case for the south transept having been built with a wooden roof is unequivocal (ref. 17). On the inner plane of the clerestory there the tall minor arches that flank the window arches have been filled in and are masked by the vault springers (Figs. 4 & 6). The vault of the clerestory passage is about 11 feet from the floor and takes no account of abutment for the high vault. The south transept high vault ribs are enriched with chevron ornament as in the nave aisle vaults from bay 3 to the west and the nave high vault (Fig. 7) (1128-1133). They are typologically later than the north transept high vault ribs which have the same profile as the choir and transept aisle vaults. For the north transept

the evidence speaks of the intention to build the high vault from the first and that this scheme was carried out as planned.

12. The wall arch of the north transept terminal wall is carried on a shaft that rises from the ground and courses consistently with the Romanesque stonework of the main arcade and gallery (Figs 3 & 8). Set on the gallery sill next to this shaft, another shaft carries the north-east diagonal rib of the high vault. This arrangement recalls the junction of the eastern crossing arch and the western bay of the choir, where the shaft from the ground carries the outer order of the crossing arch while the shaft at gallery level receives the diagonal rib of the high vault. Therefore, if the choir was initially designed for a high vault then so was the north transept. This is confirmed by the south-east vault shaft and capital of the north transept high vault which courses with the crossing pier just as the gallery sill-set shafts at the west end of the choir (Fig. 9). In each case they were intended to carry the ogives of the high vault. And they contrast with the clumsy setting of the inserted high-vault capital in the north-east angle of the south transept. The bed-joints to the left of the ribs between bays 1 and 2 on the east side of the north transept do not range with those to the right (Fig. 10). If the ribs had been added to a spandrel which was formerly plain or articulated with a continuation of the paired shafts, then the coursing of the

Fig. 9. Durham Cathedral, north transept, south east high-vault capital/east respond north crossing arch

Fig. 10. Durham Cathedral, north transept, high-vault springer between bays 1E and 2E

Fig. 11. Durham Cathedral, north transept, west wall

ashlar would have been consistent throughout the entire spandrel (ref. 18). In the south transept the bed-joints to the left of the ribs range with those to the right.

13. In the north transept the congruence of the stepped rhythm of the clerestory arches with the trajectory of the high vault indicates that they were conceived together. Moreover, the barrel vault of the passage immediately above the minor arches, some 7 feet 6 inches above the passage floor, permits the loading of masonry behind the haunches of the vault.

14. The change from the choir clerestory without a wall passage to the inclusion of the wall passage in the transepts was, for Bony, another factor in the change of plan from a high vault in the choir to a wood-roof over the transepts (ref. 19). While the solid-walled clerestory in the choir may be equated with the masons' caution in the construction and abutment of a high vault, the juxtaposition of a high vault and clerestory passage was compatable for other master masons in England at this time. Between 1077 and 1088 at St Albans a high groin-vault was erected over the eastern arm with a clerestory wall passage, while between 1107 and 1115 the presbytery of Hereford Cathedral was built with a

high groin or rib-vault and a clerestory wall passage (ref. 20).

15. The treatment of bays 1 and 2 of the west wall of the north transept is seemingly more problematic to seeing the high vault as an integral part of a single design (Figs 3 & 11). Had the high vault been part of the original scheme then why are the ribs carried on corbels rather than on attached shafts? The west side of the transept differs from the east side in not having an aisle and gallery; in place of a gallery there is a triforium. Inclusion of attached shafts for the diagonals of the high vault would have proved impractical in the west wall, given the necessity of the triforium passage to provide access from the stair turret in the north-west angle of the transept to the north nave gallery. The triforium passage is 25 inches wide. The sill on the east gallery on which the vault shafts sit is 13 inches deep. If vault shafts been employed on the west wall the 13 inch set back would have resulted in a completely useless 12-inch wall passage (ref. 21). The regular coursing of the south-west corbel and its abacus with the north-west crossing capitals and the adjacent ashlar of the north transept west wall, proves that this corbel is not a later insertion (Fig. 12). This is in contrast to the north-west corbel of the south transept which does not even align with the south-

Fig. 12. Detail of Fig. 11, corbel next west respond north crossing arch

Fig. 13. Christchurch Priory, choir crypt, west wall

west crossing capitals (Fig. 6). While the slight irregularities in the setting of the other corbels on the west wall of the north transept may suggest that the corbels and vault springers were additions, the snug fit of the regular ashlar courses and the clerestory strings against the ribs indicates that wall and vault were conceived as one. Once again, this contrasts with the south transept where the clerestory strings and the stonework adjacent to the high vault springers show clear signs of having been hacked back to facilitate the insertion of the vault. The evidence therefore suggests that the corbels on the west wall, and the high vault of the north transept, were intended from the first.

16. In the north transept the paired shafts rising from the gallery sill between bays 3E and 4E appear to be intended to carry diagonal ribs as in bays 1E and 2E (Figs. 3 & 8). However, since bay 4E is too narrow to be vaulted as a separate unit, and given the intrusion of the stair-turret, the west wall opposite bays 3E and 4E is expressed as a single bay (ref. 22). The paired shafts rising from the gallery to the vault webbing between bays 3E and 4E were therefore never intended as rib responds, but were included to continue the symmetry in bay articulation (ref. 23). Analogous wall articulation rising to the vault web is seen in the west wall of the main crypt at Christchurch Priory (Fig. 13).

17. If the north transept was built with the high vault as planned, but the wood roof constructed over the south transept represents a change in plan, then it is possible that the north transept was built before the south. This would be unusual for a church with monastic buildings on the south side. However, delay in completing the south transept may have occurred if the terminal wall was contiguous with the Anglo-Saxon cathedral which remained in part until St Cuthbert was translated in 1104 (ref. 24).

THE NAVE
18. Bilson's analysis of the nave has demonstrated that the first great campaign terminated after the easternmost double bay of the main arcade and the accompanying aisles, and one bay of the gallery on both sides (Fig. 7) (ref. 25). In this phase nothing of the clerestory was constructed. The most obvious difference between the two phases is the introduction of chevron ornament in phase 2. In the diagonal ribs of the aisle and high vaults it flanks the torus roll in preference to the simple hollows in phase 1, and it used on the second order of the transverse arches. It adorns the main arcade arches, the enclosing arches of the gallery and the arches fronting the windows in the clerestory. The high vault has pointed transverse arches in contrast to the semi-circular ones in the transept and presumably in the choir. As in the transepts there are no transverse ribs between the bays above the minor piers and the ribs are carried on corbels as on the west wall of the transepts, rather than on shafts starting at the gallery sill as in the choir and the east side of the transepts. In contrast to the setting back of the choir and transept galleries, the front of the nave gallery is in the same plane as the main arcade. This was not a change that took place in phase 2 but was set in the easternmost gallery bay in the first great campaign of construction. Other changes are also evident between the nave aisles and the aisle bays to the east. In the aisles of the choir and transepts the backs of the minor piers have three stepped shafts on dosserets to receive the ribs and transverse arch of the vault. The same form was used in the responds. In the nave the diameter of the minor piers is increased from six to seven feet and the stepped shafts at the back are abandoned, the arched elements of the vault springing instead from the back of the cylinder. This cylindrical form is then mirrored in the semi-cylindrical responds. The nave aisles are also wider than either the choir or

transept aisles. The choir aisles are 15 feet 3 inches in width as opposed to 17 feet 4 inches in the south nave aisle and 17 feet 5 inches in the north nave aisle.

19. The absence of sill-set vault shafts in the nave gallery suggests that the construction of the nave high vault was not part of the original plan. This is confirmed by the responds of the west crossing arch, in the relative alignment of the clerestory and gallery bays, and in the spandrels between the gallery arches. With one exception, the articulation of the crossing conforms to the logical arrangement of one shaft with one capital for one order of the arch. The exception is the outer order of the western crossing arch which is cut by the eastern diagonals of the nave high vault (Fig. 14). Clearly the shafts of the responds which now carry these diagonal ribs were originally intended to receive the outer order of the western crossing arch. At this stage a wood-roof was planned for the nave (ref. 26).

20. When work resumed on the nave, the plan for a wood-roof was abandoned in favour a high vault. With the front plane of the nave gallery set in the easternmost bay in line with the main arcade, to revert in subsequent bays to the recessed gallery front that had been used in the choir and transepts would have been aesthetically inappropriate. Therefore, in the absence of sill-set shafts to carry the diagonal ribs of the high vault, the ribs rest on grotesque head corbels of the same type as used in the west wall of the south transept (ref. 27). The pointed transverse arches of the nave high vault are traditionally seen as structurally superior to semicircular transverse arches in that they avoid the flattening at the crown associated with a round-headed arch over such a wide span (ref. 28). Whether this was a serious consideration for the Durham master is a moot point. It may well be that, like later English medieval architects, he was concerned to keep the tops of the transversals and diagonals at approximately the same level, and the pointed arch facilitated this without stilting the transversals or unduly flattening the diagonals.

21. The stepped rhythm of the clerestory arches conforms to the lunettes of the high vault, and the use of lower barrel vaults behind the haunches of the high vault achieves appropriate loading (ref. 29). The clerestory bays are not aligned with the gallery bays but rather with the vault lunettes. Next to the shafts of the major piers, the diagonal ribs are set on corbels next to the capitals of the transverse arch, that is, just outside the line

Fig. 14. Durham Cathedral, west crossing arch, detail north springer

above the capital of the outer order of the gallery arch. The intermediate vault corbels are set above the true centre of each minor gallery pier, that is inside the line of the capitals of the gallery arch outer order. The centre of the lunette of the vault and the clerestory is therefore slightly more towards the centre of the double bay than the gallery arches. Concomitantly, the vault is built with the clerestory. The masonry in the nave gallery spandrels which does not range from one side of the high vault ribs to the other confirms that the vault was built with the clerestory. Had the ribs been added to a formerly plain spandrel then the ashlar would have coursed consistently through the whole spandrel (ref. 30).

ABUTMENT OF THE HIGH VAULTS

22. It is generally assumed that quadrant arches across the nave galleries provide abutment for the high vault and as such are forerunners of flying buttresses (ref. 31). The proto-flying buttress theory has been questioned by Conant and Grodecki, and most specifically by Stephen Gardner (ref. 32). These quadrant arches originally comprised only a single order of voussoirs just 15-inches thick (ref. 33). Like the diaphragm arches in the choir and transepts they joined the gutteral wall level with the high-vault capitals. Quite apart from their diminutive scale compared with the 7-feet-thick wall, the quadrants at no time had masonry above them to abut the haunches of the high vault (ref. 34). It is clear, then, that the nave gallery quadrants were not conceived as buttresses for the high vault but rather as supports for a novel roof system with transverse gables, just as Gardner suggested.

SHORTENED TITLES USED

CURRY, Aspects - CURRY I. Aspects of the Anglo-Norman Design of Durham Cathedral. Archaeologia Aeliana, 1986, 5th Series, vol. 14, 31-48.

HOEY, Pier Form - HOEY L. Pier Form and Vertical Wall Articulation in English Romanesque Architecture. JSAH, 1989, vol. 48, 258-283.

PEERS et al, Report - PEERS C.R., BILSON J. and BRAKSPEAR H. A report to the Society of Antiquaries on certain repairs now being undertaken at Durham Cathedral. Proceedings of the Society of Antiquaries, 1915-16, 2nd Series, vol. 28, 49-53.

PEVSNER N. The Buildings of England: County Durham 2nd rev. edn., ed. E. Williamson, Harmondsworth, 1983.

THURLBY M. The Romanesque and Early Gothic Fabric of Durham Cathedral, in Durham Cathedral: A Celebration, ed. D. Pocock. Durham, 1993.

REFERENCES
1. SNAPE M.G. Documentary Evidence for the Building of Durham Cathedral and Its Monastic Buildings. Medieval Art and Architecture at Durham Cathedral: BAA CT for the year 1977 (1980), 20-36 at 24.
2. HILLS G.M. The Cathedral and Monastery of St Cuthbert at Durham. JBAA, 1866, vol. 22, 197-237 at 203; BILSON, Chronology, 123-128.
3. cf. HOEY, Pier Form, 269 n.71.
4. JAMES J. The Rib Vaults of Durham Cathedral. Gesta, 1983, vol. 22, 135-139.
5. VCH Durham, 103-106; PEVSNER, Durham, 183; HELIOT P. Du Carolingien au Gothique: L'evolution de plastique murale dans l'architecture religieuse du nord-ouest de l'europe (IXe-XIIIe siecle), 25 n.5. Paris, 1966; CURRY, Aspects, 32-34.
6. CLARK W.W. The Nave Vaults of Noyon Cathedral. JSAH, 1977, vol. 36, 30-33.
7. BILSON J. Recent Discoveries at the East End of Durham Cathedral. Arch. J., 1896, vol. 103, 5-7.
8. The evidence for towers over the eastern bays of the aisles is discussed by THURLBY, Durham.
9. BILSON, Durham, fig. 4.
10. ANDREW M. Chichester Cathedral: the Problem of the Romanesque Choir Vault. JBAA, 1982, vol, 135, 11-22, unsuccessfully proposes a high barrel-vault for the Romanesque choir of Chichester Cathedral.
11. On the shift from vertical in the Durham walls, see PEERS et al, Report, 50-53.
12. BILSON, Beginnings, 312-314; BILSON, Durham, 110-113, 128-140; BONY J. Le projet premier de Durham: Voutement partiel ou voutement total? in Urbanisme et architecture. Etudes ecrites et publiees en l'honneur Pierre Lavedan, 41-49. Paris, 1954.
13. On the west wall of the transept, BILSON, Beginnings, 312, expressed the firm opinion that the lack of shafts to receive the ribs of the vault indicated the temporary abandonment of the plan to vault. However, in BILSON, Durham, 110, he is less emphatic: "The omission of the vaulting shafts in the triforium stage may foreshadow the abandonment of vaulting, though such lapses from strict logic are very common in England." He believed that the paired shafts between bays 3 and 4 on the east side of the north transept that start at the gallery sill and continue up to the webbing of the vault, were originally intended for diagonal ribs as in the first double bay of the transept (BILSON, Beginnings, 313; BILSON, Durham, 136).

Bilson's chronology is followed by: VCH Durham, 110-111; CLAPHAM A.W. English Romanesque Architecture After the Conquest, p. 39; without detailed discussion of the north transept. Oxford, 1934; CURRY, Aspects, 35-36.

14. Bony is followed by PEVSNER, Durham, 170, 183-185; and PEVSNER N. and METCALFE P. The Cathedrals of England: Midland, eastern and Northern England, pp. 84-86. Penguin, Harmondsworth, 1985. FERNIE E. The Romanesque Church of Waltham Abbey. JBAA, 1985, vol. 138, 48-78. HOEY, Pier Form, 269. KIDSON P [et al. A History of English Architecture, p. 50. Penguin, Harmondsworth.], observes that "Between the choir on the one hand and the nave and transepts on the other there are slight but significant differences which seem to imply that the decision to vault the rest of the church was an afterthought."

15. The length of the choir major pier is 208 inches, measured at the top of the plinth, as opposed to 125 inches for the transept major pier. The application of the choir-size pier to the transept would consequently reduce the width of bay 2E from 90 to 48 1/2 inches, and of bay 3E from 85 to 43 1/2 inches.

16. For Gloucester nave see WILSON, Serlo's Gloucester, 53, 75 n.5; THURLBY, Tewkesbury and Gloucester, 47-48; for Malmesbury, WILSON, Serlo's Gloucester, 82 n.96. BILSON, Beginnings, 306, believed that the Lindisfarne nave was rib-vaulted and this opinion was followed by THOMPSON A.H. Lindisfarne Priory, pp. 12-14. HMSO, London, 1975. McALEER J.P. The Upper Nave Elevation and High Vaults of Lindisfarne Priory. Durham Archaeological Journal, 1986, vol. 2, 43-53, contends that the late 18th-century watercolours, sketches and sepia wash views of the Lindisfarne nave, along with the existing fabric, provide "incontrovertable evidence" that the nave was groin-vaulted. However, the NW springer of the nave high vault provides *primary* evidence for the reconstruction of a rib-vault.

17. BILSON, Durham, 128-133.

18. BILSON, Durham, 150, pl. XII, observes similar irregularities in coursing in the nave.

19. On the original form of the Romanesque choir clerestory windows see CURRY, Aspects, 34, fig. 2.

20. VCH Hertfordshire, 484, 490; WILSON, Serlo's Gloucester, 60, 78, n.43, who further suggests, 78, n.49, that the choir at St-Etienne at Caen combined a high groin-vault and a clerestory wall passage. THURLBY M. The Former Romanesque High Vault in the Presbytery of Hereford Cathedral. JSAH, 1988, vol. 47, 185-189.

21. The outer two inches of this sill is provided by the gallery string course; the front plane of the gallery wall is set back 11 inches from the front plane of the main arcade (BILSON, Durham, 125).
22. The junction of the high-vault rib with the stair turret in the Durham transepts, which seems uncomfortable to 20th-century eyes, does not appear to have concerned the Romanesque designer. It was repeated in the transepts and nave aisles at Lindisfarne Priory, and with the addition of a shaft at the angle of the turret in the western bays of the nave aisles at Durham. On the analogy of these large stair turrets with military architecture, see THURLBY, Durham.
23. HODGES C.C. Durham Cathedral. The Builder, 1893, vol 64, 427-432 at 429, suggests that the shafts between bays 3E and 4E were included "to preserve as far as possible the continuity of the design."
24. On the Anglo-Saxon Cathedral see BRIGGS H.G., CAMBRIDGE E. AND BAILEY R.N. A new approach to church archaeology: Dowsing, excavation and documentary work at Woodhorn, Ponteland and the pre-Norman Cathedral at Durham. Archaeologia Aeliana, 1983, series 5, vol 11, 79-100 at 91-97.
25. BILSON, Durham, 141-159.
26. The juxtaposition of a wood-roofed nave with a vaulted choir and transepts is found, for example at Ste-Croix at Loudun (Vienne) [CROZET R. Eglises romanes a deambulatoire entre Loire et Gironde. Bulletin Monumental, 1936, vol. 95, 61-73], Mont-St-Michel, and St-Severin-en-Condroz [BARREL I ALTET X. Belgique Romane, pp. 299-303. Zodiaque, La Pierre-qui-Vire (Yonne), 1989.
27. The use of corbels to carry vault ribs, or shafts supporting ribs, became common practice in Romanesque architecture in Britain. Out of a total of some 100 British Romanesque buildings with rib-vaults, corbels are used in nearly half. Examples occur at: Avington (Berks), chancel; Beaudesert (Warks), chancel; Berkswell (Warks), crypt; Birkenhead Priory (Cheshire), chapter house; Blewbury (Berks), chancel and tower; Blockley (Worcs), chancel; Boothby Pagnell (Lincs), Manor House, undercroft; Brabourne (Kent), chancel; Bristol, St Augustine (Cathedral); chapter house and gatehouse; Buildwas Abbey (Salop), chancel, transept chapels, slype, parlor, chapter house; Byland Abbey (Yorks), parlor; Cambridge, Holy Sepulchre, nave and aisles; Castle Rising (Norfolk), crossing; Castle Rising (Norfolk), Castle, gatehouse; Chester Cathedral, western slype; Christon (Somerset) tower; Compton Martin (Somerset), chancel; Conisborough

(Yorks), Castle Chapel; Dalmeny (Lothian), apse and choir; Elkstone (Glos), chancel; Ewenny (Glamorgan), chancel; Hampnett (Glos), chancel; Heddon-on-the-Wall (Northumberland), chancel; Hemel Hempstead (Herts), chapel; Icklesham (Sussex), tower; Kirkstall Abbey (Yorks), chancel, nave aisles, chapter house, dormitory stair, parlor, refectory undercroft, lay brothers range; Leonard Stanley (Glos), chancel; Leuchars (Fife), apse; Lilleshall Abbey (Salop), choir and transepts; Lincoln Cathedral, penultimate west bay of nave; Lindisfarne Priory (Northumberland), choir and transepts; Middleham (Yorks), Castle keep, west cellar; Newark (Notts), crypt; Newcastle Castle Chapel; Oxford Cathedral, choir aisles; Peterborough Cathedral, nave (intended); Rievaulx Abbey (Yorks), chapter house, parlor; Romsey Abbey (Hants), choir aisles, north nave aisle; Rudford (Glos), chancel; Sherborne School (Dorset), library; Sidbury (Devon), tower; Southwell Minster (Notts), nave aisles; Stow (Lincs), chancel; Warwick, St Mary, crypt; York Minster, crypt.
28. BILSON, Durham, 152.
29. BILSON, Durham, 145-146. The height of the vault is at 10' 2" behind the minor arches of the clerestory, but just 8' 5" behind the springers of the high vault.
30. BILSON, Durham, 150. The recessing of the rib springers within a pocket in the wall is followed in the undercroft of Middleham Castle (N. Yorks.), in the parlor at Rievaulx Abbey and in the dormitory undercroft at Byland Abbey, all seemingly under the influence of Durham.
31. BILLINGS R.W. Architectural Illustrations and description of the Cathedral Church at Durham, p. 5. London, 1843); MOORE C.H. Development and Character of Gothic Architecture, p. 14. Macmillan, London & New York, 1890; VCH Durham, 114; BOND F. Gothic Architecture in England, p. 370. Batsford, London, 1905; MOORE C.H. The Mediaeval Church Architecture of England, p. 35. Macmillan, New York, 1912; BOND F. An Introduction to English Church Architecture, p. 400, 403. OUP, London, 1913; PEERS et al, Report, 50-53; LEFEVRE-PONTALIS E. L'origine des arc-boutants. Congres archeologique, 1919, vol 82, 367-396, at 372-374; BILSON, Durham, 143-145; GALL E. Die Gotische Baukunst in Frankreich und Deutschland I, p. 84. Leipzig, 1925; AUBERT M. Les plus anciennes croisees d'ogives. Leur role dans la construction. Bulletin Monumental, 1934, vol. 93, 6-67 and 137-237, at 27; FOCILLON H. Art of the West in the Middle Ages, II, Gothic, p. 12. Phaidon, London, 1963; KUNSTLER G (ed.). Romanesque Art in Europe,

p. 66. Norton, New York, 1973; KIDSON P et al. A History of English Architecture, p. 52. Penguin, Harmondsworth, 1979; WILSON C. The Gothic Cathedral, p. 19. Thames & Hudson, London, 1990.
32. CONANT K. Carolingian and Romanesque Architecture 800-1200, p. 461. Penguin, Harmondsworth, 1978; GRODECKI L. Gothic Architecture, p. 39. Abrams, New York, 1977; 39; GARDNER S. The Nave Galleries of Durham Cathedral. Art Bulletin, 1982, vol. 64, 564-579; BONY J. French Gothic Architecture of the 12th and 13th Centuries, p. 485 n.10. University of California, Berkeley, 1983.
33. On the controversy surrounding the 1915 addition of two orders beneath the original quadrant arches, see PEERS et al, Report, 50-53; BILSON, Durham, 127, 129, 143; CURRY, Aspects, 34.
33. Buttresses of high groin-vaults in major Roman buildings support at the haunches of the vault; for example, Rome, Basilica of Constantine (Maxentius), well-illustrated in FLETCHER B. A History of Architecture, 19th edn, ed. J. Musgrove. London, 1987.

The purpose of the rib in the Romanesque vaults of Durham Cathedral

M. THURLBY, PhD, FSA, Associate Professor, Department of Visual Arts, York University, Ontario

SYNOPSIS
1. This paper argues the following: that the ribs were introduced at Durham primarily for aesthetic reasons; that the aisle vaults were erected on full centering just like groin-vaults, while in the high vaults the ribs served as an aid to the centering; and that the rib-vault was not considered structurally superior to the groin-vault. The sources of the Durham vaults are also examined.

INTRODUCTION
2. The publications of John Bilson have assured the Romanesque cathedral of St Cuthbert at Durham (1093-1133) a prominent place in the history of european architecture (ref. 1). This fame is largely the result of a rationalist approach which saw in embryo at Durham three motifs deemed fundamental to the Gothic system of building: the pointed arch, the rib-vault and the flying buttress (ref. 2). The validity of this linear argument is suspect. As early as 1912 it was challenged by Charles Moore for whom Durham had "no significance in connection with the beginnings of Gothic architecture (ref. 3). He continues, "Stupendous as it is, no part of Durham Cathedral has, in my judgment, any tendency in a Gothic direction. The semicircular groin rib is a feature of organic Romanesque, and the pointed arch of Durham is not used in what I consider a Gothic way - that is to say, no structural advantage is gained by it that the round arch would not give." Aside from structural concerns, Webb emphasised the "all-over linear pattern" of Romanesque Durham, an aesthetic concept embraced by Kidson and Bony, but to this day many aspects of the rationalist argument are upheld (ref. 4).

WHY WERE RIBS INTRODUCED?
3. This question has intrigued architectural historians ever since Viollet-le-Duc pronounced the

Fig. 1. Durham Cathedral, north choir aisle to east

structural properties of the rib (Fig. 1). The debate, which is conveniently summarised by Frankl down to 1960, ranges around three basic possibilities (ref. 5). First, if in a groin-vault the forces pass through the web along the shortest line to the groin and are channeled along the groin to the corners of the bay, then the reinforcement of the rubble-built groin with an ashlar rib would be structurally beneficial. Secondly, the rib could be constructionally advantageous in providing a permanent scaffold from which to erect wooden centering for the webbing in each severy, and/or a neat edge to ragged rubble at the groins. It also facilitates the creation of straight diagonals rather than the serpentine line often found in groin-vaults (Fig. 2) (ref. 6). This is allied to the third consideration, that the rib was an aesthetic device which completed the 'logical' articulation of pier and respond shafts and tied them together in the manner of a moulded arch.

4. The studies of Robert Mark and his associates have demonstrated that the forces in both groin and rib-vaults funnel in a cone-like manner towards the springers of the vault and therefore the rib-vault does not have any structural advantage over the groin-vault (ref. 7). However, as Mark admits, this does not solve the problem of what the medieval mason thought of the rib. To try to answer this we have to rely on empirical observation at Durham itself and elsewhere (ref. 8).

THE PURPOSE OF THE RIB: AESTHETIC

5. The evidence at Durham and in other in Romanesque buildings in Britain suggests that the rib was primarily conceived as an aesthetic device integral with the shafted piers and responds, and moulded arches. It gives a bold plastic expression to the diagonal of the quadripartite vault which was formerly marked by the linear groin. The painted linear articulation of vaults was not new. Roman vaults were often painted with linear geometric divisions (ref. 9). In the presbytery vault at San Vitale, Ravenna (526-547), mosaic ribbons decorate the groins (ref. 10). The tradition continues through the early middle ages in the west, as in the crypt of Saint-Germain at Auxerre, where painted ribbons adorn arches and groins, and illusionistically painted shafts and capitals articulate flat responds (ref. 11). These are 2-dimensional precursors of the fully 3-dimensional articulation at Durham. The painted shaft and capital evolve through the crypt piers at Auxerre Cathedral (ca. 1030), to the stepped aisle-vault responds at St-

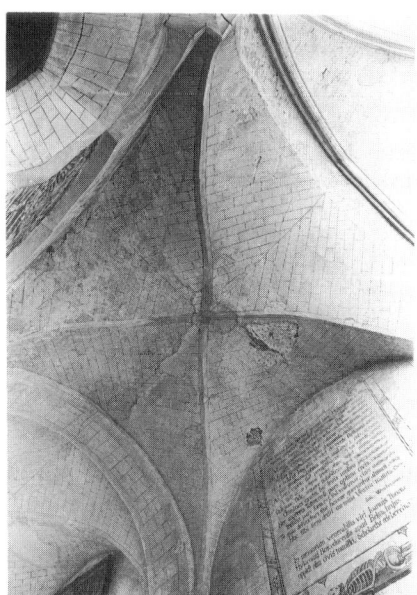 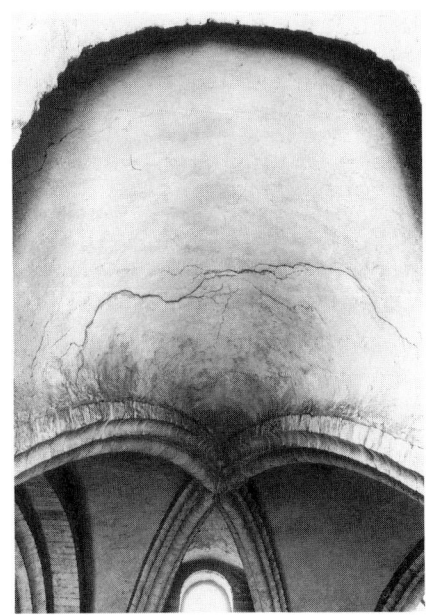

Fig. 2. St. Albans Abbey, south choir aisle vault

Fig. 3. Durham Cathedral, north transept high vault from west clerestory

Etienne at Caen, Winchester Cathedral and elsewhere, where they carry the plain transverse arch and the groins, and thence to Durham where they carry moulded transverse arch and ribs (ref. 12). Similarly, the Romanesque painted choir-aisle vaults at St Albans have the transverse arches and groins edged with ribbons (Fig. 2). These are transformed into moulded transverse arches and ribs at Durham (Fig. 1). An analogous typological evolution is apparent from painted 'marble' main arcade columns and the dado arcade arches, as at St-Savin-sur-Gartempe (Vienne), to the carved columnar piers and the arcade mouldings at Durham (ref. 13).

WERE THE RIBS USED AS A CONSTRUCTIONAL AID?
6. Bilson reports that in the nave high vault "between the tops of the ogives and doubleaux there is always a wide joint (of 2 inches or so) which received the boards of the centering on which the cells were built, and some fragments of oak boards were found in the course of repairs" (ref. 14). The impression created is of a large-scale version of the boards still in situ in a room off the stairvice in the north-east transept of Lincoln Cathedral (1192-1200) (ref. 15). Here the ribs act as permanent supports for the centering. Something

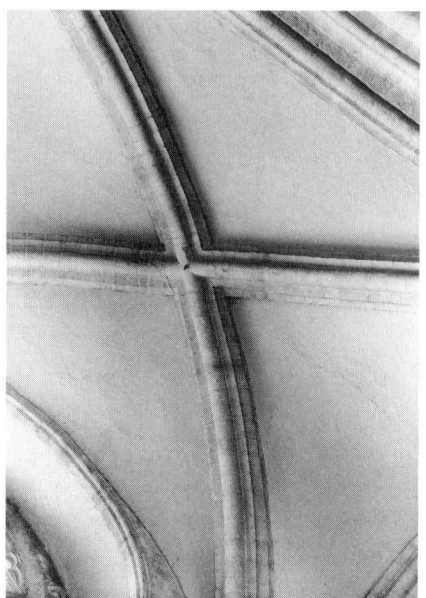

Fig. 4. Durham Cathedral, north choir aisle vault, bay 2

Fig. 5. Durham Cathedral, north choir aisle, bay 3, detail of arc beneath vault web

similar is seen in the choir vault at Stow (Lincs.) as rebuilt by Pearson (ref. 16). Here the vault springers sit directly on the ribs but after some 30 degrees there is a gap between the extrados of the rib and the vault web in which the wooden centering was placed. In certain cases the fragments of the centering remain.

7. How do these observations fit with Durham? Thin wooden centering boards are still in situ in the enclosing arches of some bays of the nave galleries. In the high vaults there are irregular ledges between the extrados of the ribs and the webbing which are presumably the remains of the now-plastered 2-inch gaps observed by Bilson. From these there is no difficulty in envisioning wooden centering planks for the longitudinal sections of webbing. But for the lateral severies that arch up from the clerestory walls, the plough-shared sections could not have been created solely on wooden planks (Fig. 3). A more maleable alternative must be sought, and here the Lincoln centering is again instructive in that it uses wattle in addition to wooden planks. Thus it is plausible to suggest that the ploughshared severies of the Durham high vaults were erected on wattle centering secured to a wooden frame that rested on the extrados of the ribs.

8. In the aisle vaults the absence of an appreci-

able gap between the extrados of the ribs and the webbing precludes the use of the ribs as a support for the centering. The vaults would have been constructed on full centering like groin-vaults. In such construction the ribs would have been supported within a cradle as suggested by Fitchen (ref. 17). The serpentine trajectory of some of the aisle ribs, especially bays 2 and 3 of the north choir aisle and bay 5 of the south nave aisle (counting from the crossing in each case), confirms that the ribs were laid up with the rubble web (Fig. 4) (ref. 18).

WERE THE RIBS AT DURHAM INTENDED FROM THE FIRST?
9. Bilson's studies have determined that the rib-vaults of the choir aisles belong to the original fabric and that a Romanesque choir high vault was replaced with the present one in connection with the construction of the Nine Altars. One detail not reported by Bilson bears on our understanding of the aisle vaults. From certain capitals carrying the diagonal ribs there is an upward arcing break in the ashlar of the wall below the junction with the vault web (Fig. 5). The arc starts below the rib and therefore cannot result from the rebuilding of a vault severy when the present windows were inserted. Instead it represents a planned lower trajectory for the original vault. Therefore, in spite of their apparent sophistication, the vaults were not achieved without constructional difficulties. The Durham master mason experienced problems analogous to, albeit less extreme than, those in the crypt at Lastingham (1078-85) where the initial arced termination of the aisle walls falls well below the actual webbing of the vault. The Durham arcs seem too low to be equated with the torus profile of the present ribs, and therefore groin-vaults may have been planned the choir aisles. Groins would be better suited to the backs of the minor piers where the springers of the rib, transverse arch and outer order of the main arcade overlap (Fig. 1). Against this, the evidence for the arcs below the present vault webs is not the same in all bays which suggests that the geometry of the rib-vault was imperfectly understood. This was certainly the case in the presbytery at Dalmeny (Lothian) where similar arcs appear on the side walls beneath the web of the rib-vault.

ANALOGIES FOR THE AESTHETIC USE OF THE RIB
10. The apses of the transept crypts at Christchurch Priory are divided into three by half shafts which continue as ribs in the vault. The 'nave' of each crypt by contrast has a barrel-vault and broad

rubble, transverse arches. Had the rib been regarded as structurally superior then it is difficult to explain why it was not used throughout the crypts. Instead the ribs are used iconographically to signify the altar bay. Similarly, in the eastern arm of Saint-Martin at Boscherville (Seine-Maritime) the apse is rib-vaulted while the two straight bays of the choir are groined (ref. 19). In the rectangular eastern arm of Ewenny Priory (Glamorgan) the eastern bay which houses the high altar is covered with a quadripartite rib-vault while the other bays are barreled (ref. 20). Then at Cormac's Chapel, Cashel (Co. Tipperary), the rectangular eastern arm is rib-vaulted while the nave is barreled (ref. 21).

11. At Worcester Cathedral the ribs of the early 12th-century chapter house are applied as pure decoration on the vault between the groins (ref. 22). In the refectory undercroft there (c. 1150-60) five-part groin-vaults cover each bay. Had the rib been considered structurally beneficial it would have been essential in this very situation to carry the superstructure of the refectory. Conversely, for this relatively unimportant space it was decided to build a rubble groin-vault to avoid the expense of ashlar ribs. In a number of 12th-century monastic houses in England rib-vaults are used selectively for the areas deemed more important, while the lesser spaces would be groined. Thus at Kirkstall, the presbytery, nave aisles, chapter house, parlour, west range are rib-vaulted while the undercrofts of the dormitory and refectory are groined (ref. 23). At Rievaulx, the chapter house and parlour were ribbed while the dormitory undercroft was groined. At Byland, the aisles throughout the church, the chapter house and parlour were rib-vaulted, the undercrofts of the dormitory and refectory were groined. At Furness rib-vaults were used in the transept chapels and the nave aisles but in the west range there were groins. At Chester, the western slype and the chapel of St Anselm immediately above it are rib-vaulted while the remainder of the west range was groined.

12. Elsewhere, groin-vaults continue to be built in major buildings well after the introduction of the rib, as in East Anglia where groins are used throughout the aisles at Ely Cathedral, Norwich Cathedral, Castle Acre Priory and Binham Priory. Construction on each of these buildings continued after the documented use of rib-vaults in East Anglia in 1118 in the aisles of Peterborough Cathedral (ref. 24). No structural benefit was seen in the rib and therefore the groin-vault continued in use in these buildings for the sake of aesthetic

uniformity. Probably under the influence of the East Anglian churches groins were used in the nave aisles at Waltham Abbey (ref. 25). The same prolonged use of groin-vaults is also witnessed in the Augustinian abbey churches of St Botolph's, Colchester, and St Bartholomew's, Smithfield. In the midlands groin-vaults were used in the choir and ground storey of the westblock at Melbourne (Derbs) and in the nave aisles of the former Benedictine priory at Tutbury (Staffs).

SOURCES OF THE DURHAM VAULTS

13. The accomplished form of the Durham vaults suggests that even if they were the first to have ribs they probably evolve from a tradition of major vault construction in post-Conquest England (ref. 26). This tradition is difficult to reconstruct given the serious losses amongst these buildings. The ribs in the apses of the transept crypts and south transept chapel at Christchurch Priory and the south transept chapel at Tewkesbury Abbey may pre-date the Durham vaults, but ribs were not used consistently elsewhere in those buildings (ref. 27). Similarly, St Peter's Abbey (now Cathedral), Gloucester, probably had a rib-vaulted apse but with either groins or a barrel over the straight bays of the choir (ref. 28). However, in the Gloucester nave the triforium is set back in the manner of the Durham choir galleries and this feature may belong to the original 1089 design (ref. 29). Bilson suggested that the choir of Romanesque Lincoln was groin-vaulted and it is virtually certain that the eastern arm of St Albans was so covered (ref. 30). St Albans is especially relevant because, like Durham, it had a four-bay eastern arm that terminated in an apse, it housed an important local saint, and was conceived on the huge scale of the great early Christian basilicas (ref. 31). Certain details of St Albans's daughter house at Tynemouth are closely allied to Durham, but unfortunately the ruined state of the church there only reveals that the nave aisles were groin-vaulted (ref. 32). Most significant is the groin-vault of the former presbytery of Lastingham Priory (Fig. 6). Although usually dismissed as Pearson's invention, documentation demonstrates that he restored the former high groin-vault (ref. 33). The plinths, responds and certain capitals at Lastingham presage those at Durham and suggest a link between the two buildings, although whether this was direct or through St Mary's Abbey, York, which was constructed by the Lastingham monks after they abandoned Lastingham, is a moot point (ref. 34).

Fig. 6. Lastingham Priory, interior to east

* For this and the preceding paper, I should like to thank The Dean and Chapter of Durham for granting me unlimited access to the cathedral. Mr Owen Rees, Head Verger, and his assistant, Mr Reg Wright, have been models of helpfulness during my many visits. Mr Ian Curry, architect to the fabric of Durham Cathedral, has accompanied me over the building on numerous occasions; kindly informed me of the erection of, and allowed me access to, the scaffolding in the choir in 1991; and has freely shared his profound knowledge of the fabric. I have also benefitted from discussions with Dr Eric Cambridge, Ms Linda Denesiuk, Mr Stuart Harrison, Professor M.F. Hearn, Professor Lawrence Hoey, Mr Hugh McCague and most particularly with Professor Eric Fernie. Financial support was provided by the Social Sciences and Humanities Research Council of Canada.

SHORTENED TITLES USED

Arch. J. - Archaeological Journal.

BAA CT - British Archaeological Association Conference Transactions.

BILSON, Beginnings - BILSON J. The Beginnings of Gothic Architecture. II. Norman Vaulting in England. JRIBA, 1899, vol. 6 289-319, discussion, 319-326.

BILSON, Chronology - BILSON J. Durham Cathedral: The Chronology of Its Vaults. Arch. J., 1922, vol. 79, 101-160.

JBAA - Journal of the British Archaeological Association.

JRIBA - Journal of the Royal Institute of British Architects

JSAH - Journal of the Society of Architectural Historians.

THURLBY, Tewkesbury and Gloucester - THURLBY M. The Elevations of the Romanesque Abbey Churches of St. Mary at Tewkesbury and St. Peter at Gloucester. Medieval Art and Architecture at Gloucester and Tewkesbury: BAA CT for the year 1981 (1985), 36-51.

THURLBY, Ewenny - THURLBY M. The Romanesque Priory Church of St. Michael at Ewenny. JSAH, 1988, vol 47, 281-294.

TRANS. RIBA. - Transactions of the Royal Institute of British Architects.

VCH - Victoria History of the Counties of England.

WILLIS, English Cathedrals - WILLIS R. Architectural History of Some English Cathedrals, vol. 2. Minet, Chicheley, 1973.

WILSON, Serlo's Gloucester - WILSON C. Abbot Serlo's Church at Gloucester (1089-1100): Its Place in Romanesque Architecture. Medieval Art and Architecture at Gloucester and Tewkesbury: BAA CT for the year 1981 (1985), 52-83.

REFERENCES
1. BILSON, Beginnings; BILSON, Chronology.
2. FOCILLON H. The Art of the West in the Middle Ages, I, Romanesque Art, pp. 53-54. ed. J. Bony. Phaidon, Ithaca, N.Y., 1980.
3. MOORE C. The Mediaeval Church Architecture of England, p. 25. Macmillan, New York, 1912.
4. WEBB G. Architecture in Britain: The Middle Ages, pp. 38-39. Penguin, Harmondsworth, 1956; KIDSON P et al. A History of English Architecture, p. 53. Penguin, Harmondsworth, 1979; BONY J. Durham et la tradition saxonne. Etudes d'art medievale offertes a Louis Grodecki, 79-85 at 79-80. Paris, 1981. See also BOASE T.S.R. English Art 1100-1216, p. 21. OUP, Oxford, 1953.

Perpetuation of the rationalist view of Durham is often found in texts used for first-year university courses in art history, for example, DE LA CROIX, H et al, Gardner's Art Through the Ages, p. 358. 9th edn, Harcourt Brace Jovanovich, San Diego, 1991.
5. FRANKL P. The Gothic: Literary Sources and Interpretations through Eight Centuries, pp. 663-666, 763-772, 798-826. Princeton, 1960.
6. On the geometry of groin-vaults, see BILSON, Beginnings, 293-295, fig. 6.
7. ALEXANDER K.D., MARK R. AND ABEL J.F. The Structural Behavior of Medieval Ribbed Vaulting. JSAH, 1977, vol. 36, 241-251; MARK R. Experiments in Gothic Structure, pp. 102-117. M.I.T., Cambridge [Mass.], 1982.
8. cf. WILLIS R. On the Construction of the Vaults of the Middle Ages. Trans. RIBA, Part 2, 1842, vol. 1, 1-63 at 3, reprinted, WILLIS, English Cathedrals.
9. JOYCE H. The Decoration of Walls, Ceilings, and Floors in Italy in the Second and Third Centuries. G. Bretschneider, Rome, 1981.
10. KRAUTHEIMER R. Early Christian and Byzantine Architecture, ill. 190. 4th edn. rev. by R. Krautheimer and S. Curcic. Penguin, Harmondsworth, 1986.
11. HUBERT J. et al. The Carolingian Renaissance, ill. 5. Braziller, New York, 1970.
12. For Auxerre Cathedral crypt, see CONANT K. Carolingian and Romanesque Architecture 800-1200, ill 112. 2nd integrated edn (revised), Penguin, Harmondsworth, 1978. For stepped, shafted responds at St-Etienne at Caen, Winchester Cathedral and elsewhere, see BILSON, Beginnings, 293, figs. 2 & 4.
13. DEMUS O. Romanesque Mural Painting, ill. 29. Abrams, New York 1970.
14. BILSON, Chronology, 156.
15. WILSON C. The Gothic Cathedral, ill. 17. Thames & Hudson, London, 1990.

16. ATKINSON G. On the Restorations in Progress at Stow Church, Lincolnshire. Associated Architectural Reports and Papers, 1850-1, vol. 1, 315-326; SPURRELL M. Stow Church Restored 1846-1866. Lincoln Records Society, 1983, vol 75.
17. FITCHEN J. The Construction of Gothic Cathedrals, figs. 59 & 60. Chicago, 1977.
18. JAMES J. The Rib Vaults of Durham Cathedral. Gesta, 1983, vol. 22, 135-145 at 139-140. James's "slightly exaggerated" sketch of the buckling of the ribs (139, fig. 8) is grossly exaggerated.
19. WILSON, Serlo's Gloucester, pl. XI F.
20. THURLBY, Ewenny, fig. 3.
21. STALLEY R. Three Irish Buildings with West Country Origins. Medieval Art and Architecture at Wells and Glastonbury, BAA CT for 1978 (1981), pl. XIV C.
22. WILLIS R. The Crypt and Chapter House of Worcester Cathedral. Trans. RIBA, 1862-3, 1st Ser., 13, 213-225, reprinted in WILLIS, English Cathedrals, figs. 11a & 13.
23. ST JOHN HOPE W. H. and BILSON J. Architectural Description of Kirkstall Abbey. Leeds, 1907.
24. On Peterborough Cathedral aisle vaults, see BILSON, Beginnings, 302-305.
25. FERNIE E. The Romanesque Church of Waltham Abbey. JBAA, 1985, vol. 138, 48-78.
26. McGEE D. The 'Early Vaults' of Saint-Etienne at Beauvais. JSAH, 1984, vol. 45, 20-31. On ribs from Saint-Lucien at Beauvais, commenced c. 1090, see HENRIET J. Saint-Lucien de Beauvais: mythe ou realite. Bulletin Monumental, 1983, vol. 141, 273-294.
27. Christchurch Priory was commenced by Ranulf Flambard after 1087, see HASE P.H. The Mother Churches of Hampshire, 45-66 esp. 58-60, in Minsters and Parish Churches: the Local Church in Transition 950-1200, ed. John Blair. Oxford, 1988. Evidence remains for groin-vaults in the nave aisles and beneath the former transept galleries at Christchurch. Tewkesbury was commenced by Robert Fitzhamon after 1087 [DUGDALE W. Monasticon Anglicanum, II, p. 53. London, 1819]. It was covered throughout with high barrel vaults, had groin-vaulted choir aisles, quadrant vaulted nave aisles, and may have had a rib-vaulted apse, see THURLBY M. The Romanesque Elevations of Tewkesbury and Pershore. JSAH, 1985, vol. 44, 5-17. For the apse rib-vaults at Christchurch and Tewkesbury, and stimulating comments on apse ribs in general, see WILSON, Serlo's Gloucester, 64-66. 28. For different interpretations of the former Romanesque choir high vaults at Gloucester, see WILSON, Serlo's Gloucester

(groin-vault), THURLBY, Tewkesbury and Gloucester (barrel), and THURLBY, Ewenny. Wilson also reconstructs a high groin-vault in the choir of Tewkesbury.
29. WILSON, Serlo's Gloucester, 53, 75n.5; THURLBY, Tewkesbury and Gloucester, 47-48.
30. BILSON J. The Plan of the First Cathedral of Lincoln. Archaeologia, 1911, vol. 62, 543-564 at 558; PEERS C.R. and PAGE W. VCH Hertfordshire, II, p. 484, 490; WILSON, Serlo's Gloucester, 60, 78n.43.
31. FERNIE E. The Effect of the Conquest on Norman Architectural Patronage. Anglo-Norman Studies, 1986, vol. 9, 71-85 at 85, refers to the "near megalomania" of the new patrons, and suggests [Observations on the Norman Plan of Ely Cathedral, Medieval Art and Architecture at Ely Cathedral: BAA CT for the year 1976 (1979), 1-7 at 4] "that the great length of the more important Anglo-Norman churches is an attempt to emulate the size of the largest Early Christian basilicas in Rome." GEM R. The Romanesque Cathedral of Winchester: Patron and Design in the Eleventh Century. Medieval Art and Architecture at Winchester Cathedral: BAA CT for the year 1980 (1983), 1-12, compares the scale of the Romanesque cathedral with St Peter's, Rome.
32. On Tynemouth Priory see KNOWLES W.H. The Priory Church of St Mary and St Oswin, Tynemouth, Northumberland. Arch. J., 1910, vol 67, 1-50.
33. QUINEY A. John Loughborough Pearson, p. 132. Yale U.P., New Haven and London, describes the vault as "a complete falsification." The present high vaults at Lastingham date from Pearson's restoration of 1879, but a citation and faculty of 1877, preserved in the Borthwick Institute, University of York (FAC 1877/2a, FAC 1877/2b), in which the language is very precise, refer to the proposal to "restore the ancient groining of the nave (i.e. the square presbytery bay) in place of the present plaster ceiling," and to "restore the ancient barrel vaulting of the chancel including the Apsidal East End." I should like to thank Dr Christopher Norton for suggesting that I contact the Borthwick Institute in connection with documentation on Lastingham.

Durham Cathedral tower vibrations during bell-ringing

J. M. WILSON and A. R. SELBY, University of Durham

ABSTRACT. The main central tower of Durham cathedral houses a ring of ten bells of varying size and mass. The heaviest, tenor, bell has a mass of 1400 kg. The bells are supported on bearings and each can be swung through a vertical arc of $360°$, some in a North-South direction and others in the East-West plane. When a bell swings full circle it acts as a compound pendulum and exerts significant dynamic forces, both horizontal and vertical, on the bell frame and thence on the tower structure. During change ringing the bells are rung in complex sequences known as 'methods' and intricate patterns of forces act on the tower which are the aggregate of those for each of the bells.

Measurements of the tower movements have been taken to determine the maximum velocities in the N-S and E-W directions, the natural frequencies of tower oscillations and the damping factors in response to various modes of bell ringing.

The transient dynamic response of the tower was estimated by a combination of computational exercises. Finite element models of the tower were constructed using beam elements and 3-dimensional brick elements. These models were calibrated for natural frequency against measured values. A separate program was used to compute bell forces acting on the tower at regular time intervals. Finally, the transient response of the beam model to these forces was calculated.

The measured and computed responses were found to be in reasonable overall agreement.

INTRODUCTION.
The Tower Structure
1. Construction of the Cathedral was begun in 1093 and the present structure with the central tower belfry was essentially completed by 1490. The foundations rest directly on to sandstone bedrock, except at the west end, and the construction material is sandstone, mostly quarried from across the river. Structurally the tower comprises four massive clustered columns leading up to semi-circular Norman arches. Above, the walls of the tower

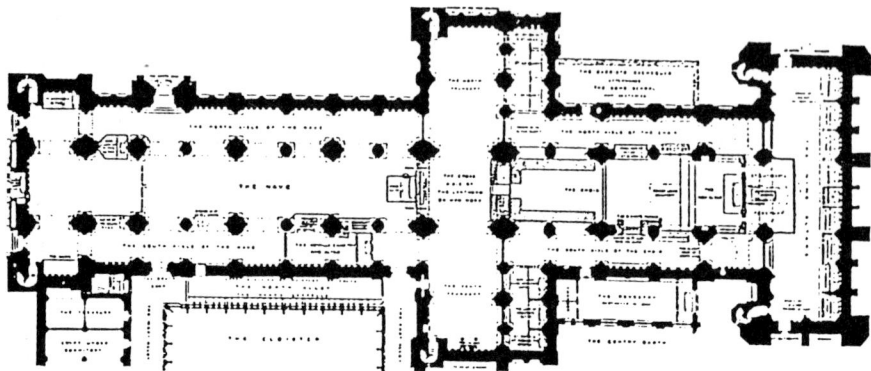

Figure 1. Plan of Durham Cathedral.

Figure 2. Vertical section through the tower.

are thickwall construction, comprising inner and outer skins of dressed stone with tie-stones, and a rubble infill. Additionally, the corners of the tower are thickened and buttressed which compensates in part for the large fenestrations and other cut-outs, and the accommodation of a spiral staircase in the south-west corner of the tower. The tower is some 66m in overall height, with the bell frame being at a height of about 60m above foundation level. The nave, choir and transepts form a cross structure in plan view, and each arm of the 'cross' frames into the tower structure at levels below some 24m above ground (Figures 1 & 2).

The Ring of Bells

2. Durham Cathedral has a fine ring of ten bells hung in the main centre tower. Their voices can be heard over wide distances in the neighbourhood. Each bell is bolted to a headstock at one end of which is attached a lightweight timber-spoked rope wheel from which hangs the bell rope. The headstock is supported by two trunnion bearings which in turn rest on a steel bell frame (Figure 3). The bell frame is securely fixed to the masonry of the tower, and each bell is mounted to swing in either an N-S or an E-W direction. The orientation and position of the bells within the bell frame is such as to produce a circular arrangement of the bell ropes. The bells vary in size and mass and have been installed at various dates. The first ring to be hung in the main tower, in 1693, comprised five bells, which were derived by recasting some smaller bells which had previously hung in the northwest tower. Bells were added in 1781, 1896 and then in 1980. The bells were all cast in the Whitechapel foundry. The masses of the bells, their dates of casting, and their notes, in the key of D major, are given in Table 1.

Table 1 Bell data.

Bell No.	Mass(kg)	Date	Note
1 (treble)	330	1980	F#
2	359	1980	E
3	389	1780	D
4	400	1693	C#
5	507	1781	B
6	563	1896	A
7	639	1693	G
8	801	1693	F#
9	1096	1693	E
10 (tenor)	1425	1693	D

Until 1980 the eight bells were supported by a timber frame, but this had shrunk and warped over many years, and it was replaced by a frame of

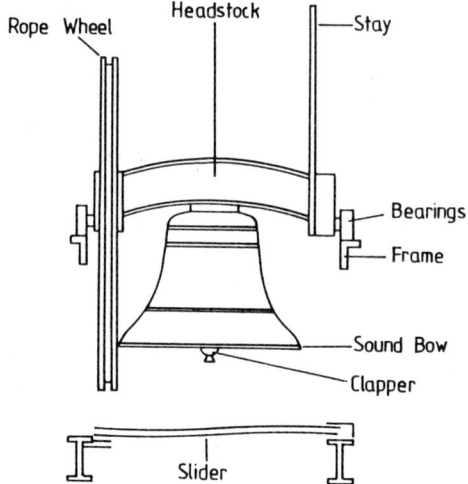

Figure 3. Bell with ringing fixtures and supports.

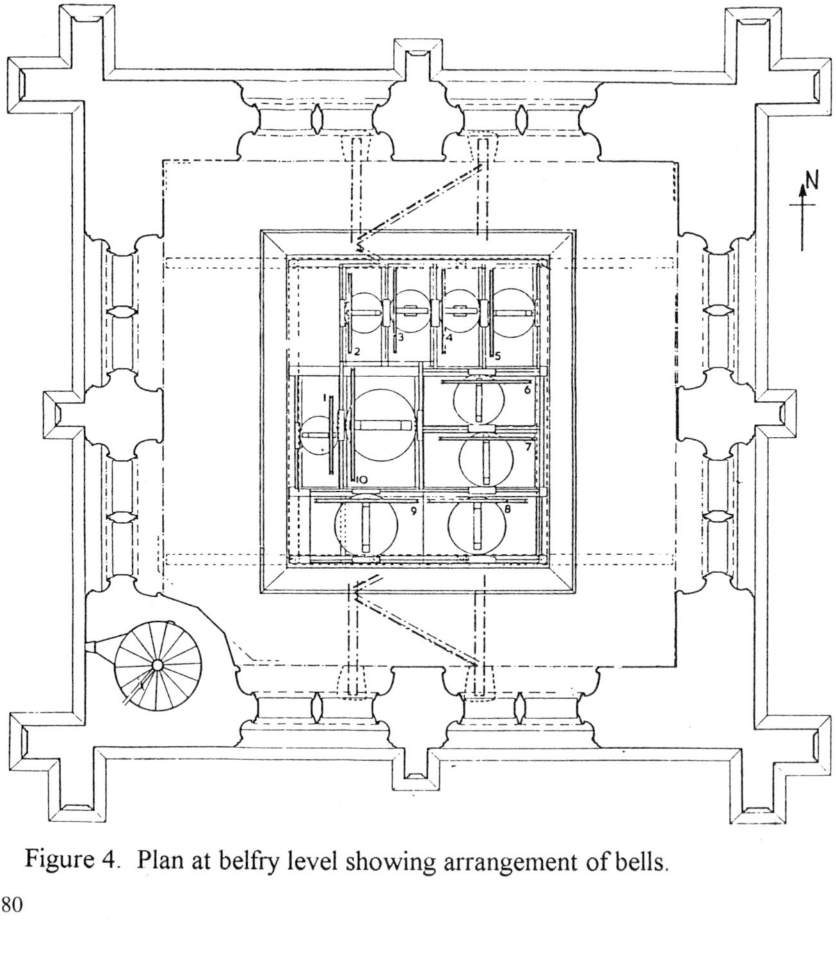

Figure 4. Plan at belfry level showing arrangement of bells.

structural steel components well fixed into the masonry of the tower (Figure 4).

Forces due to Bell Ringing

3. The bells are rung by swinging them through a full circle from the mouth-up position, in the English system. The ringer initiates the swing of the bell by pulling the bell rope (Figure 5). When a bell swings freely about an axis which does not coincide with its centroid of mass, it acts as a compound pendulum (Figure 6). The action of the swinging bell transmits forces through the bell frame to the tower which depend on the mass and shape of the bell, and on the degree of eccentricity of the centroid from the axis of suspension. A number of investigators have published work relating to the analysis of bell forces. Heywood (1914) made some important advances in the early days of bell tower analysis, and a significant theoretical study was published by Lewis (1914) which estimated forces produced by the ringing of bells. Lewis also showed that the forces produced during the full circle rotation were approximately four times the bell weight in the vertical direction and some twice the bell weight in the horizontal. More recently, Heyman and Threlfall (1976) found good agreement between theoretical predictions of inertial forces produced by bell ringing and those determined from laboratory tests. They also developed a useful method for evaluating the inertial properties of a bell from simple in-situ measurements, which are described later. The maximum horizontal force is typically of the order of twice the static weight of the bell and headstock. It reverses in direction during one full swing of the bell and tends to cause sway of the tower. The maximum vertical force, which occurs as the bell swings through the mouth-down position is typically of the order of four times the static bell weight and acts downwards adding to compressive forces in the tower.

4. The arrangement of bells within the bell frame serves to reduce the likelihood of excessive horizontal forces and torques being applied to the tower. Thus as far as possible the weights of bells swinging in opposite and orthogonal directions are matched and the larger bells are placed as close as possible to the tower axis.

The Present Study

5. In this current study the response of the tower is investigated by a combination of measurements and of analysis. The reason for this is that neither method on its own gives a complete picture of tower behaviour.

6. It is possible to take measurements of the bells in-situ to determine their moments of inertia and mass eccentricities. This information, together with the masses recorded by the foundry, allows detailed analysis to be made of the forces applied to the tower by each swinging bell. These forces may then be incorporated in a dynamic numerical analysis of the tower

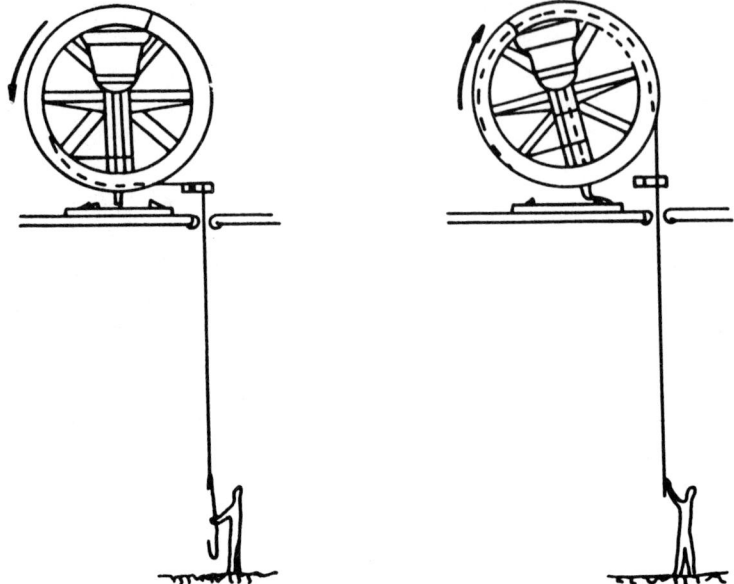

Figure 5. English system of bell ringing.

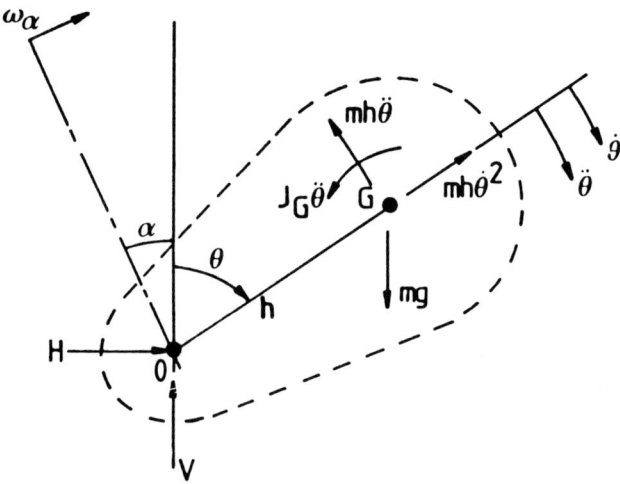

Figure 6. Dynamic equilibrium of compound pendulum.

which can be represented either as a vertical cantilever beam or by a more realistic three-dimensional finite element model. However there are still some uncertainties in the parameters affecting the analysis, including the stiffness of the masonry walls with arches and cut-outs, the unknown stiffness contributions of the nave and transept structures and the level of damping.

7. To measure both the characteristic natural frequencies and mode shapes of free vibration, together with their corresponding levels of damping, or the response to forced vibration of any large structure requires the use of a large number of low-frequency high sensitivity motion transducers and associated multi-channel data recording and processing equipment. Such equipment was not available for this study.

8. However with comparatively few transducers it is possible to identify the natural frequencies and levels of damping present in a few dominant modes of vibration. With some discrete measurements of tower vibration combined with a computational study it was hoped that a fuller picture of the behaviour of the tower could be obtained.

EVALUATION OF BELL FORCES.
Theory

9. A bell may be considered as a compound pendulum the behaviour of which is characterised not only by its mass (as in the case of a simple pendulum) but also by its moment of inertia about its centroid. Figure 6 shows a compound pendulum pivoting about a fixed point, O. The centroid of mass, m, is at point G, and the radius of gyration of the bell is k, such that the moment of inertia of the bell about its centroid is J_G (where $J_G = mk^2$). The moment of inertia of the bell about point O, J_O, at distance h from the centroid is given by

$$J_O = J_G + mh^2 = m(k^2 + h^2) \tag{1}$$

For small oscillations, the frequency of oscillation, ω is given by

$$\omega^2 = \frac{gh}{(h^2 + k^2)} \tag{2}$$

corresponding to a period, τ, of

$$\tau = 2\pi \sqrt{\frac{h^2+k^2}{gh}} \tag{3}$$

10. Next, consider full-circle rotation of the bell, ignoring any losses from friction and assuming that the bell is given a small initial angular velocity

(ω_α in Figure 6) or impulse by the ringer causing it to move from the dwell position ($\theta = -\alpha$ in Figure 6) through the mouth-up position.

11. Introduction of a parameter c, an inertial form factor (Heyman & Threlfall 1976) where

$$c = \frac{h^2}{h^2 + k^2} \tag{4a}$$

and of p, a parameter slightly in excess of unity (depending upon the initial velocity and dwell position) where

$$p = \cos \alpha + \frac{\omega_\alpha^2 \tau^2}{8\pi^2} \tag{4b}$$

gives an expression for vertical force V, as a function of the general bell position, θ, of

$$V = mgc \left(\frac{1-c}{c} + 3 \cos^2\theta - 2p \cos\theta \right) \tag{5}$$

12. If the initial velocity is assumed to be so small that p is unity, then the maximum value of vertical force V is simply

$$V_{max} = mg(1 + 4c). \tag{6}$$

13. The horizontal force H is given by

$$H = mgc \sin\theta \, (3 \cos\theta - 2p) \tag{7}$$

If p is again taken to be unity, then the above expression takes a maximum value, at $\theta = 124°$, of

$$H_{max} = 3.05 \, mgc. \tag{8}$$

Traces of the horizontal and vertical forces generated by the ringing of the tenor bell are shown in Figure 7, as a function of angular position.

14. The previous analysis relates the dynamic forces to the angular **position** of the bell during its swing. However, it is necessary to extend the analysis to derive the forces as functions of **time** which are then in the form required for dynamic analysis of the tower structure.

15. The initial angular velocity can no longer be approximated to zero, as this would lead to quite unrealistic cycle times. To determine the time taken

TOWER VIBRATIONS

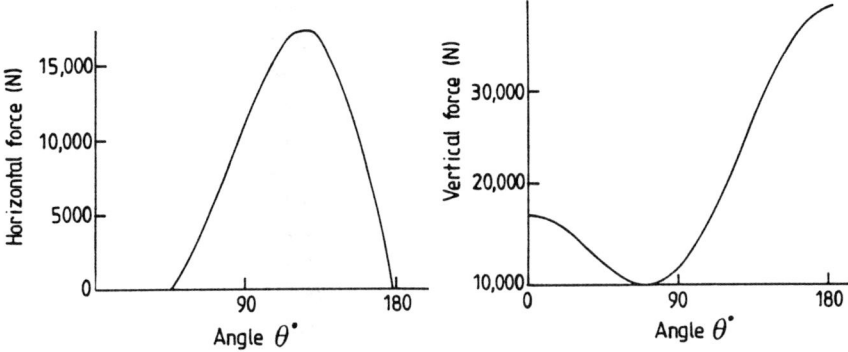

Figure 7. Horizontal and vertical forces from bell 10, as a function of position.

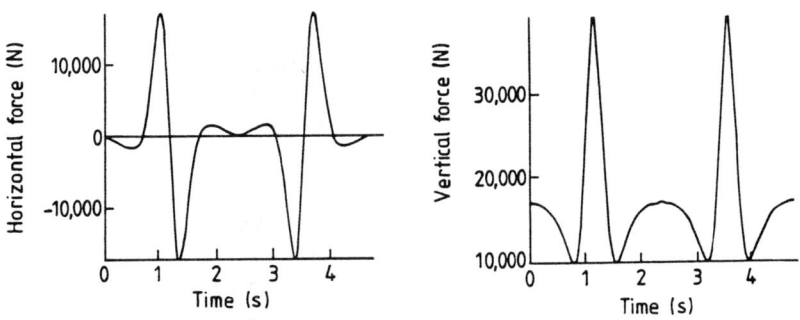

Figure 8. Horizontal and vertical forces for bell 10, as a function of time.

for the bell to move from the dwell position (where $\theta = -\alpha$) to a general position θ, an elliptic integral of the first kind must be introduced, the value of which depends on α, θ and ω_α. Elliptic integrals of this form can be evaluated analytically for trivial cases only. In general they must be computed using a numerical procedure, for example that developed by Birlisch (1965) and implemented in standard computer algorithms. Conversely, to find the angular position of the bell after the elapse of a given time within the swing, an inverse elliptic function, the Jacobian, must be used. Again these are found as standard computational algorithms. Hence the forces can be calculated at regular intervals of time, by means of equations (5) and (7).

16. For certain simple methods of ringing, such as tolling, firing, or ringing rounds, the forces acting on the tower are periodic and so can be represented by Fourier series. These can be approximated to finite series using the computational procedures described in detail by Wilson (1988).

17. The forces computed as functions of time are strongly characteristic in form as shown in Figure 8. Each trace defines the horizontal or vertical force function during a complete oscillation, i.e. a swing through an angle of ($360^o + 2\alpha$) and return, of the tenor bell. The duration of such a double revolution of a bell in the Cathedral is typically slightly less than 5 seconds (in this case 4.92 seconds). Because the bell rotates slowly near the top of its swing and rapidly through the mouth-down position, the traces are more "spiky" than those of Figure 7.

18. As an essential precursor to these detailed calculations it was necessary to deduce the mass, eccentricity and moment of inertia of each bell (including headstock and rope wheel) in the ring. This was undertaken in the following manner.

Measurement of bell characteristics.

19. The straightforward method of estimation of the mass of any bell is by reference to records kept by the foundry and often found in the church. In the case of the bells in the Cathedral, their masses are displayed in the ringing chamber of the belfry, as reproduced in Table 1. The composite mass of the bell assembly, m, which constitutes the compound pendulum, (and comprises the bell, the headstock and rope wheel) was estimated by applying a factor of 1.2 to the mass of the bell alone, in accordance with a result obtained by Heyman and Threlfall (1976).

20. The eccentricity of the bell mass centroid from the axis of rotation, h, was deduced from a simple set of observations, following the technique developed by the same authors. A mass of known weight W was attached to the bell rope and measurement was taken of the downward movement, d, of a marker on the rope, see Figure 9. Further increments of load up to 600N were applied to the bell rope, and the measurements were repeated. From static equilibrium,

TOWER VIBRATIONS

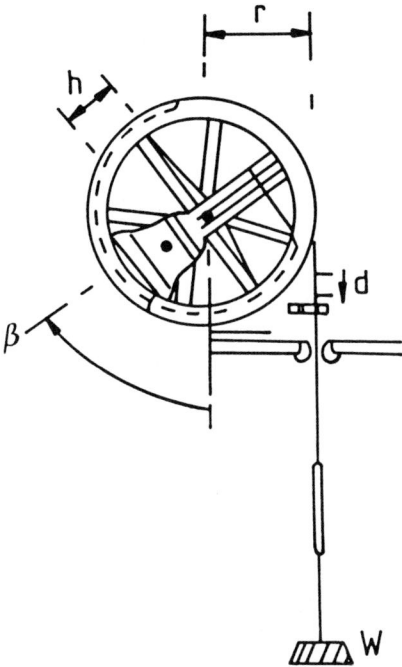

Figure 9. Measurements for position of bell centroid.

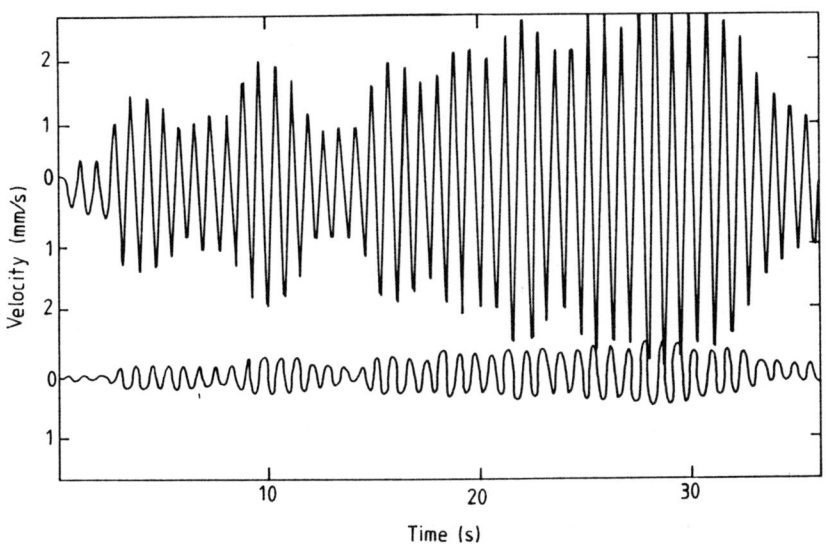

Figure 10. Measured N-S response to tolling of bell 10, at ringing chamber and at clerestory.

$$Wr = mgh \sin \beta . \qquad (9)$$

and

$$d = r\beta \qquad (10)$$

where r is the radius of the rope wheel and β is the angular movement of the bell. From a plot of W against d, the gradient can be used for calculating h. By timing the period of small oscillations of a bell, τ, the radius of gyration of the bell may be derived from equation (3).

Calculations of Bell Forces

21. These observations were taken for each of the ten bells in turn. The results obtained by direct measurement and by use of equations (3) to (10) are summarised in Table 2. Because the values of inertial form factor for the Cathedral bells ranges between 0.4 and 0.62, the maximum horizontal force applied by any single bell, as given by equation (7), is of the order of 1.5 times its own weight. The maximum vertical force applied is typically three times its own weight (see equation (6)) in addition to its static weight.

Table 2. Dynamic Characteristics of the Bells

Bell Number	Mass (kg)	Eccentricity	Radius of gyration	Small Oscillation Period	Inertial Form Factor	Maximum Dynamic Force	
						Horizontal	Vertical
		h (m)	k (m)	τ (s)	c	H_{max} (N)	V_{max} (N)
1	330	0.393	0.306	1.594	0.622	6140	11290
2	359	0.355	0.315	1.60	0.558	5993	11380
3	389	0.360	0.354	1.689	0.508	5917	11570
4	400	0.345	0.381	1.756	0.451	5394	11000
5	507	0.408	0.375	1.742	0.541	8211	15737
6	563	0.409	0.391	1.774	0.523	8796	17045
7	639	0.408	0.431	1.863	0.473	9034	18130
8	801	0.432	0.461	1.926	0.468	11227	22568
9	1096	0.426	0.497	2.013	0.424	13888	28987
10	1425	0.433	0.527	2.08	0.403	17170	36514

MEASUREMENTS OF TOWER MOVEMENT DURING BELL-RINGING.

22. An important element of a study of the bell tower response to bell ringing is an in-situ measurement programme. Small, low frequency movements of a tower can be detected by a number of systems, where the primary transducer may indicate displacement, velocity or acceleration. For example, Maunder (1981) measured the movement (displacement) of Exeter cathedral tower using a video camera trained on a target. The present authors (Selby and Wilson, 1991) used low-frequency high-sensitivity

geophones (velocity transducers) and a digital computerised data recorder and processor to measure the response of a small tower, that of St. Brandon's Church, Brancepeth. Finally Wimmer et al (1989) used piezoelectric transducers to measure accelerations of an Austrian bell tower.

23. In this study a high sensitivity low frequency Mark geophone was used to generate an analogue voltage signal proportional to the transient velocity of the tower, and the trace was displayed on an ink-pen chart recorder. Digital computer-based data recorder and processors could be used (see Selby and Swift, 1990), but these are generally better suited for much faster, short event monitoring, and do not offer immediate signal display.

24. Recordings of tower velocity were made during tolling of bells 9 and 10, and during the ringing of rounds when all ten bells in descending order were rung. For each recording, a geophone was placed on a stone shelf at the level of the ringing chamber, and connected to the chart recorder. Traces were recorded for an interval of about one minute in each of these modes, and examples are shown in Figures 10 to 13. During some recordings, a second geophone was placed lower down the tower at clerestory and simultaneous traces were obtained. At the termination of the tolling of bells 9 and 10, the recordings were continued while the tower movement decayed. These traces, shown in Figures 14 and 15, were of specific value in allowing damping factors to be deduced from the logarithmic decay, and the natural frequencies of free vibration of the tower to be estimated without disturbance caused by bell motion. The natural frequencies of the tower in the fundamental sway modes were found to be 1.31 Hz and 1.28 Hz in the N-S and E-W direction respectively. The damping ratio, in each mode, was found to be 0.016.

A summary of maximum transient velocities measured in each of the cases mentioned above is given in Table 3.

25 All the above recordings were made in terms of transient velocity, since direct measurement of displacement is more difficult and less reliable. However, because of the predominance of the fundamental sway frequency, f, in the waveforms, it is not unrealistic to make an estimate of maximum displacement, u_{max}, simply as

$$u_{max} = \frac{\dot{u}_{max}}{2\pi f} \tag{11}$$

where $\dot{u}_{max}$ is the maximum velocity.

26. A simple estimate by this device yields maximum displacements of the order of a mere 0.43 mm (equivalent to velocities of 3.5 mm/s) (see Table 3). These small values may surprise the casual observer, who can positively feel tower movement when standing in the ringing chamber during ringing. However, it is well-recognised that the human frame is highly sensitive to

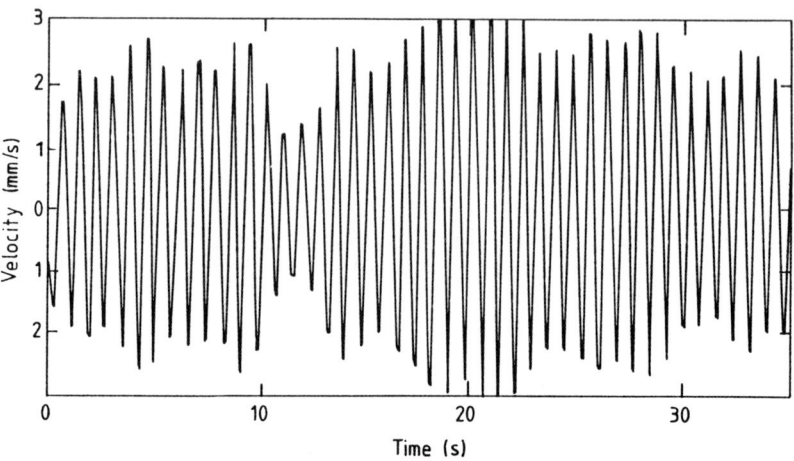

Figure 11. Measured N-S response at ringing chamber to ringing rounds on bells 1 to 10.

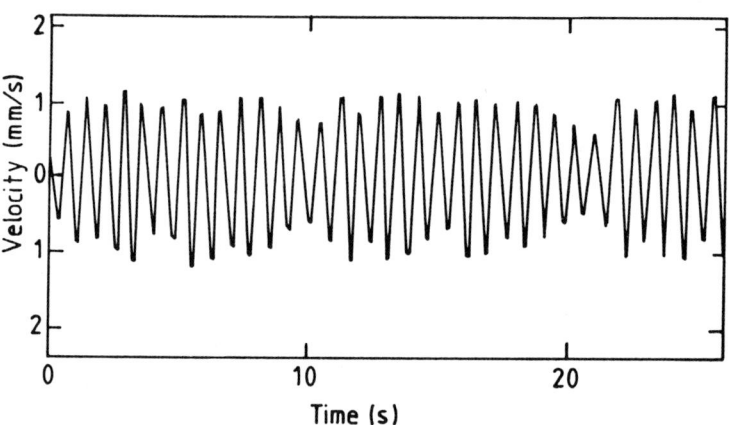

Figure 12. Measured E-W response at ringing chamber to tolling of bell 9.

TOWER VIBRATIONS

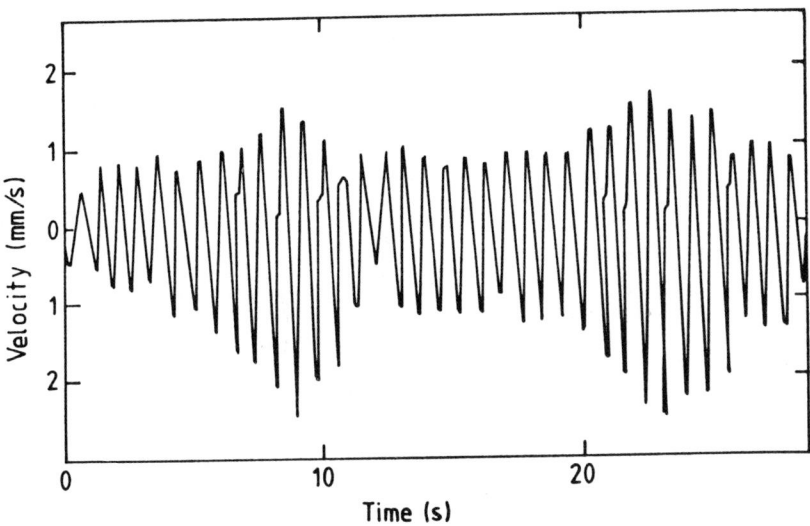

Figure 13. Measured E-W response at ringing chamber to ringing rounds on bells 1 to 10.

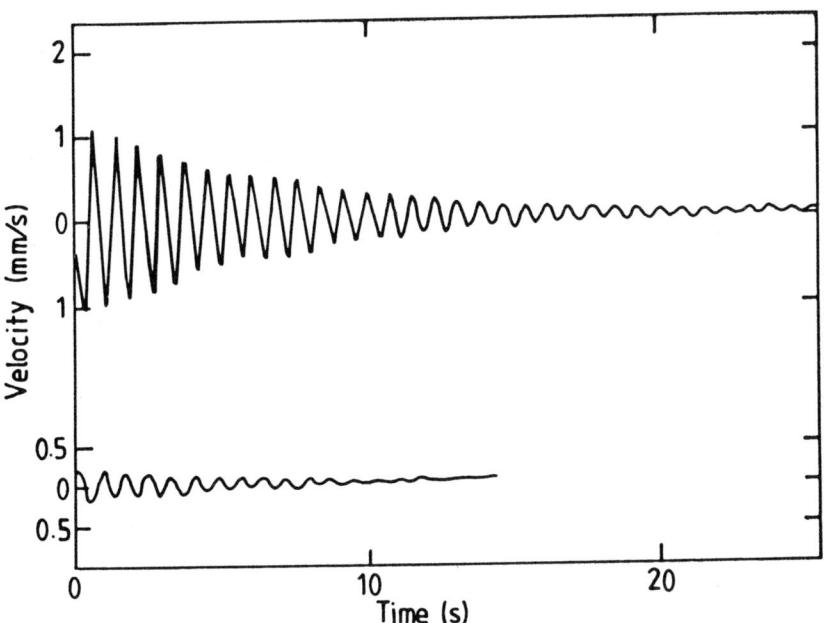

Figure 14. Decay of N-S tower vibration.

vibration and capable of detecting vibration velocities as small as 0.1 mm/s., Steffens (1985).

27. It is possible also to make estimates of dynamic strains and stresses at the base of the tower, due to bell ringing, from the estimated displacements by assuming simple cantilever bending. The results are included in Table 3, and are clearly of very small magnitude, especially in comparison with the static compressive stress due to self weight, of some 3.2 MPa.

Table 3. Measured Tower Response to Bell Ringing

Ringing Mode	Measurement Direction	Ringing Interval (s)	Peak Velocity at Ringing Chamber (mm/s)	Derived Peak Displacement (mm)	Base Strain $\times 10^{-6}$	Base Stress (MPa)
Tolling 10	N-S	2.46	2.8	0.34	3.6	0.05
Tolling 9	E-W	2.52	1.3	0.16	1.7	0.03
Rounds 1-10	N-S	2.33	3.5	0.43	4.5	0.07
Rounds 1-10	E-W	2.33	2.5	0.31	3.3	0.05

NUMERICAL MODELLING OF THE TOWER.
Introduction

28. The dynamic response of a structural system can be derived by solution of the general equation:

$$\underline{M}\,\underline{\ddot{x}} + \underline{C}\,\underline{\dot{x}} + \underline{K}\,\underline{x} = \underline{F}(t). \tag{12}$$

where $\underline{M}$, $\underline{C}$ and $\underline{K}$ are respectively the mass, damping and stiffness matrices for the structure, $\underline{\ddot{x}}$, $\underline{\dot{x}}$ and $\underline{x}$ are respectively the acceleration, velocity and displacement vectors and $\underline{F}(t)$ is a time-dependent force vector. The nature of these equations is such that they can only be solved numerically using computers. Nowadays, finite element methods, in which the structure is split up into a set of elements of simple shape and finite size, are used in computer programs to generate automatically the system equations.

29. An exact computer model of the tower is difficult to achieve because of the following aspects which are difficult to quantify.
  (i) Restraint conditions of the foundations of the tower. In the case of the Cathedral tower, the structure is founded directly on to sandstone bedrock, with considerable bearing pressure, and so fixity against both displacement and rotation is a reasonable assumption.
  (ii) Partial restraint provided by structural connections. In this case the connection details of the masonry walls of the nave, choir and transepts are not easily identified.
  (iii) The stiffness of the tower walls. Although the plan sections of the walls at critical levels up the tower are available, it is not a trivial

TOWER VIBRATIONS

exercise to apportion effective bending stiffnesses at sections including arches, fenestrations, stairs or passages, and buttresses. Also of significance is the form of the walls themselves which are of sandwich construction with outer leaves of dressed sandstone, linked by intermittent tie-stones, with a rubble fill between. As an approximation the walls can be assumed to be solid stone with an effective elastic modulus reduced, in compensation, from the typical value 17 GPa to some 6 GPa.

30. If realistic conditions and values can be ascribed, bearing in mind the above limitations, then the tower can be modelled using a finite element method, either by means of beam elements or three-dimensional solid brick elements producing a more representative structure. The quality of the finite element model can then be evaluated by comparing the computed values of the fundamental natural frequencies in the N-S and E-W sway modes with those found from the times of free vibration following cessation of ringing. When a satisfactory finite element model has been derived, then transient response may be estimated using a forward time-stepping routine.

Computer Models of the Cathedral Tower

31. A computer model of the cathedral tower, as shown in Figure 16a, was developed using 144 three dimensional twenty-noded brick elements (corresponding to roughly 3000 degrees of freedom) within the PAFEC (Henshell, 1984) suite of programs. The model, which was reasonably accurate geometrically, featured cut-outs for the belfry louvres and the large windows, and a system of springs at clerestory level to emulate the interaction of the tower at the crossing formed with the nave and transept. The columns supporting the tower were modelled, albeit rather crudely, and their bases were assumed to be fully fixed. The Young's modulus of the masonry was taken to be 5.8 GPa and the density to be 2200 kg/m^3, these figures being derived from that of the local sandstone (measured at 17 GPa and 2200 kg/m^{-3} respectively). The reduced modulus was adopted to allow for the flexibility of the masonry joints and the use of rubble infill (with much reduced structural stiffness) for the walls estimated to account for about half the nominal wall thickness of tower. The effective density was assumed to be unchanged. Where other significant reductions in section were present, for example the gallery section and arches, the Young's modulus and density of the masonry were further reduced in proportion to avoid further geometric complication of the model. The resulting natural frequencies and mode shapes computed using PAFEC for the most significant modes are shown in Table 4. It can be seen that computed fundamental N-S and E-W sway frequencies have been arranged to approximate the measured frequencies by the use of these modified material properties. The fundamental N-S sway mode is illustrated in Figure 16b, that for the E-W direction being very similar.

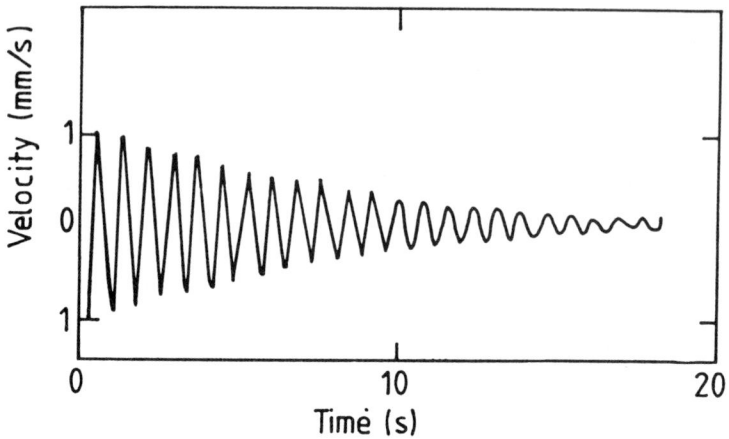

Figure 15. Decay of E-W tower vibration.

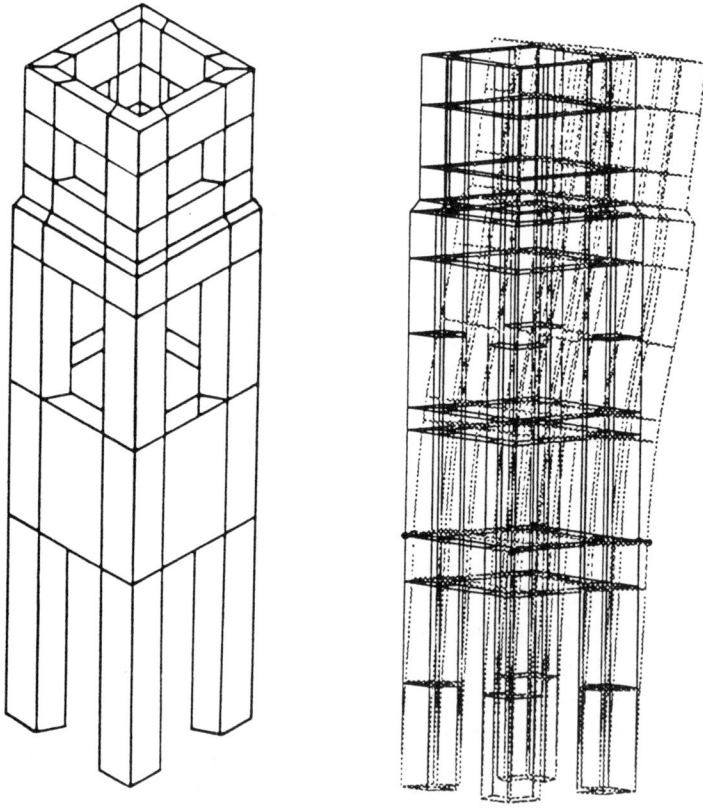

Figure 16 (a) 3D finite element model.   (b) Fundamental sway mode.

32. The ringing of any or all of the bells in a regular manner produces a set of periodic forces on the tower the response of which can also be computed using the same program suite. However the use of the brick element model to compute the response would have proved prohibitively wasteful in computational resources. Hence a simpler model was sought which retained the capability of representing the dynamic behaviour of the tower to a tolerable level of accuracy. Such a model was constructed, using special beam elements. These are capable of including shear deflection (as well as bending) and the effects of rotary inertia and are suitable for use with stubby towers. Using this model, with the same material and spring constants as for the brick model, only 54 degrees of freedom were required to produce a model having similar dynamic characteristics in terms of the computed natural frequencies and mode shapes, as shown in Table 4. If all but the sway freedoms in one particular direction (e.g. N-S or E-W) are removed, only 18 degrees of freedom are required, resulting in further opportunity for economy.

Table 4. Natural Frequencies (Hz) Calculated Using Finite Element Models

Mode No.	Element Type	Sway E-W	Sway N-S	Torsion	Bounce
1	Brick	1.29	1.30	2.71	5.94
	Beam	1.28	1.31	3.56	8.35
2	Brick	4.16	4.32	6.73	9.27
	Beam	5.40	5.52	8.62	16.8
3	Brick	7.90	7.91	7.54	-
	Beam	10.9	10.9	12.3	26.0

Computation of the Tower Response

33. Using a computer program, described by Wilson (1988) and outlined in a previous section, the forces due to bell ringing in each of the three modes indicated in Tables 3 and 5 were calculated for total ringing times of about 50 seconds. This time was sufficient to enable a steady state of vibration to be developed. Because it was found experimentally that ringing bells in these modes produced no measurable response other than sway in the directions of forcing, the economical beam finite element model was chosen to represent the tower structure. The sway responses were computed, with excitation provided by the horizontal forces only, using the Newmark β method within the PAFEC program suite. For the response calculations the parameter β was set to ¼ for stability and a time step of 0.002s was chosen to ensure convergence of the numerical solution. Output in the form of plots of velocity as a function of time was requested at positions corresponding to the ringing chamber and the clerestory levels.

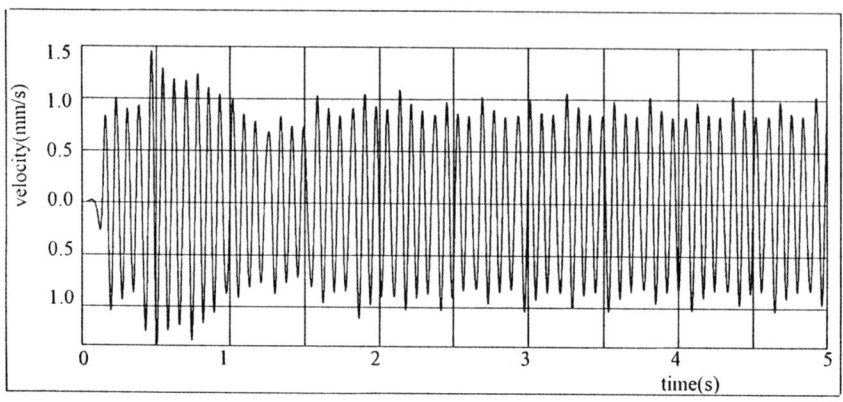

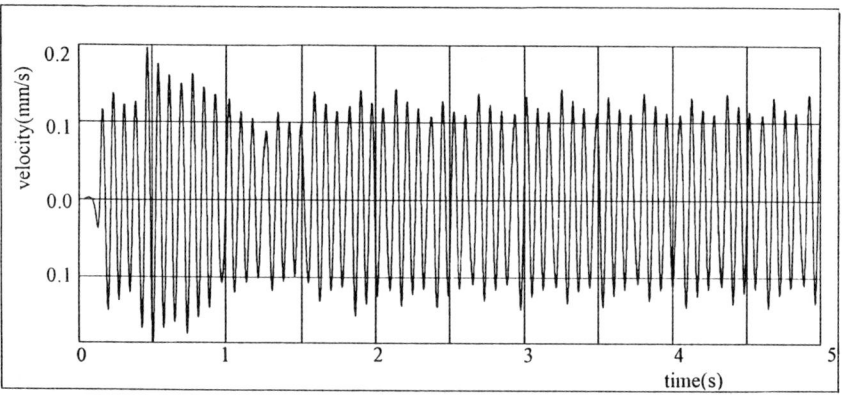

Figure 17. Computed response to tolling bell 9, at ringing chamber and clerestory level.

These are shown in Figures 17 and 18 for bells 9 and 10 respectively. The peak velocities computed and measured at these levels are shown in Table 5.

Table 5. Computed and Measured Peak Velocities (mm/s) at Ringing Chamber and Clerestory Level for Bells 9 and 10.

Ringing Mode	Measurement Direction	Ringing Interval (s)	Peak Velocities (mm/s)			
			Ringing Chamber		Clerestory	
			Measured	Computed	Measured	Computed
Tolling 10	N-S	2.46	2.8	1.3	0.6	0.23
Tolling 9	E-W	2.52	1.3	1.4	-	0.20
Rounds 1-10	N-S	2.33	3.5	1.5	-	0.30
Rounds 1-10	E-W	2.33	2.5	0.7	-	0.14

Discussion

34 It can be seen that while the results for peak velocities for tolling bell 9 agree to within 10%, those computed for tolling bell 10 and ringing rounds are only some 50% or less of the measured values. The results for bell 9 may well be fortuitous in view of the number of assumptions made in computing the response.

35. The bell forces will contain errors due to assumptions made concerning the mass of the rope wheel and fittings, the action of bell ringer in stopping and starting the ringing motion and experimental errors. Because the periodic motion of the bell produces harmonic forces having frequencies which are simple multiples of the ringing frequency, and the level of damping measured and introduced into the computation is low, the approximations incurred in deriving the computer model may have had comparatively significant effects on the magnitude of the resulting response. These approximations arise chiefly from the material properties adopted for masonry, the boundary conditions applying and the nature of the constraints provided at the tower crossing.

36. When making a detailed comparison between the computed and recorded traces, it is of interest to investigate the consequences of variation of ringing speed. By reference to the earlier analysis, it is noted that the duration of the complete cycle of a single bell from dwell through $(360 + 2\alpha)^o$, pause, and return to rest was typically about 4.8 seconds. However, the dwell-time between swings is fully controlled by the technique of the bell-ringer, and even the time-duration of the 360° swing can be strongly influenced by the angular velocity of the bell at mouth-up position as developed by the ringer. Variation in the dwell was investigated over a rather wide range of possible values, by considering the largest value of

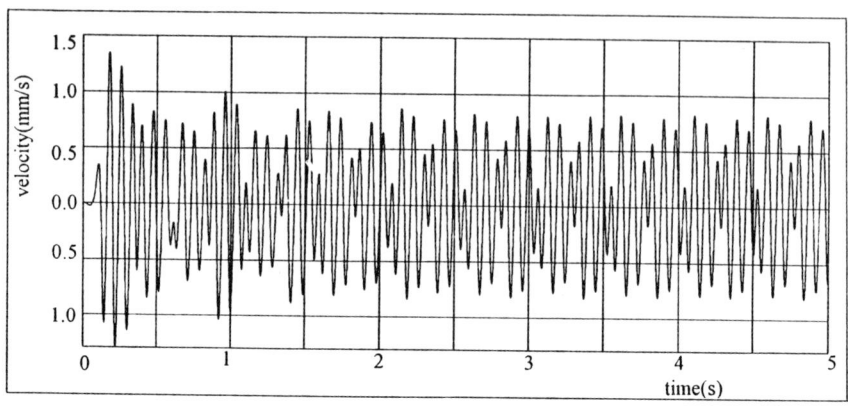

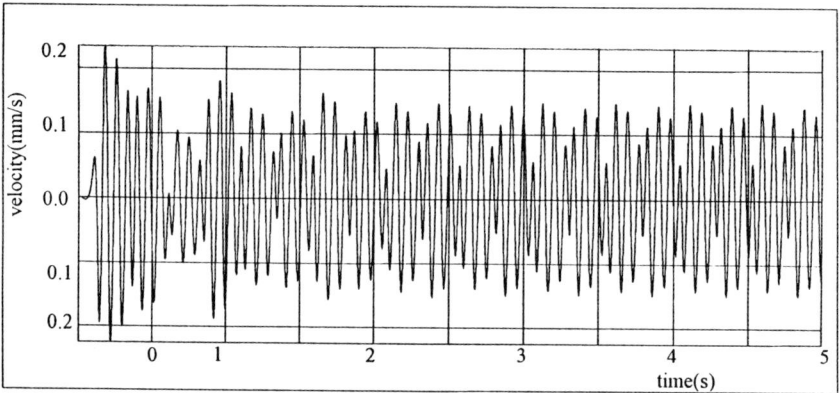

Figure 18. Computed response to tolling bell 10, at ringing chamber and clerestory level.

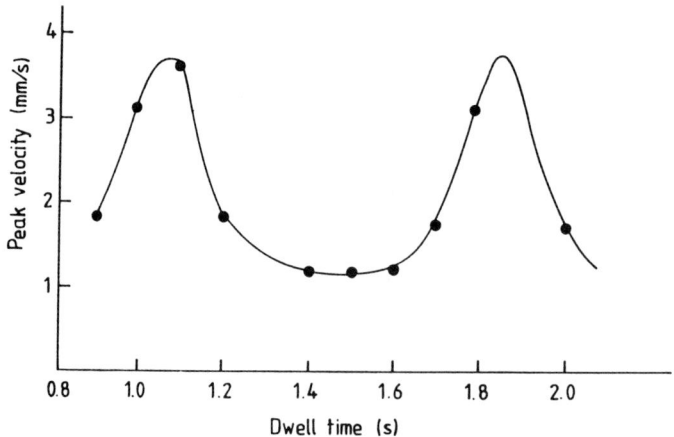

Figure 19. Response to bell 10 as a function of dwell time.

tower velocity in response to tolling of bell 10, over some five complete ringing cycles. The results show two clear peaks of response, with a flattish trough between in which the realistic value of dwell time of 1.3s lies, see Figure 19. It is probably fortuitous that the dynamic response characteristics are lowest for the excitation caused by the normal bell-ringing rate.

CONCLUSIONS

37. A detailed study of the forces exerted by the bells during ringing and of the dynamic response of the tower has been undertaken, using a combination of measurements, recordings, analysis and computation. The computed peak horizontal forces varied from 5400N for bell 4 up to 17000N for bell 10. The dynamic vertical forces were estimated to range between 11300N and 36500N. The horizontal forces exerted on to the tower caused it to sway in the manner of a vertical cantilever beam. The fundamental natural frequencies of the tower in N-S and E-W motion were measured to be 1.28 Hz and 1.31 Hz respectively. Damping of the structure was measured to give a damping ratio of 0.016.

38. The actual magnitudes of dynamic tower deformation were very small, with transient velocities generally less than 3.0 mm/s., with equivalent displacements of some 0.4 mm. Clearly, these movements cause no hazard to the tower structure. However, the integrity of the bell bearings, frame and connections to the masonry is important, so that the forces exerted by the bells can be transmitted to ground with no incremental damage at the local fixings.

39. A beam finite element model was developed which gave adequate representation of the resonant behaviour of the tower. A more sophisticated 3-D finite element model gave little improvement for considerably greater computer cost. The beam model was then used in transient dynamic analyses to compute the tower response to tolling of single bells and the ringing of rounds. Reasonable agreement was obtained with recorded tower dynamic behaviour, but it was deduced that variation in the 'dwell time' in the ringing pattern could result in significant changes in maximum response.

ACKNOWLEDGEMENTS.

40. Thanks are expressed to the following: The Dean and Chapter, Mr Curry, Cathedral Architect, Mr Trigg, the Chapter Steward, Ian Tilling and the cathedral ringers, especially Albert Bokma and Edward Cheeseman, and Richard Barratt.

REFERENCES

1. Birlisch, R. 1965, Numerical calculations of elliptic integrals and elliptic functions, Numerische Mathematik, 7, 78-90.

2. Evans, N.C. 1990, The response of bell towers to bell ringing, Final Year Honours Project, School of Engineering and Computer Science, University of Durham.

3 Henshell, R.D. (Ed) 1984, PAFEC Data Preparation User Manual, PAFEC Ltd. Strelley Hall, Nottingham.

4. Heyman, J. and Threlfall, B.D. 1976, Intertia forces due to bell ringing, International Journal of Mechanical Sciences, 18, 161-164.

5. Heywood, Sir A.P. 1914, Bell towers and bell hanging - an appeal to architects, MS in library of Central Council of Church Bell Ringers.

6. Lewis, E.H. 1914, Calculation of the forces acting upon a church tower, MS in library of Central Council of Church Bell Ringers, Penmark House, Guildford, Surrey, GU1 1BL.

7. Maunder, E.A.W. 1981, Masonary buildings in service under dynamic loading, Cement and Concrete Association, Research Seminar, University of Exeter, 6th-10th July.

8. Selby, A.R. and Wilson, J.M. 1991, The dynamic response of a church tower to bell ringing, Structural repair and maintenance of historic buildings, Vol. 2 - Proc. of 2nd Int. Conf. Seville. Comp. Mech. Publ. Southampton, ISBN 1-85312-151-7 pp3-16.

9. Steffens R.J. 1985, Structural vibration and damage, BRE Report, Watford.

10. Wilson, J.M. 1988, Periodic forces on bell towers arising from bell ringing, Internal. Report. School of Engineering and Computer Science, University of Durham.

11. Wimmer, H., Majer, J. and Niderwanger, G. 1989, Dynamic behaviour and numerical simulation of old bell towers, Structural Repair and Maintenance of Historical Buildings (Ed. C.A. Brebbia), Computational Mechanics Publications, Southampton.

Cathedral engineering for novices

J. WELFORD, AKC, PhD, Lecturer in Computer Science, University of Durham

SYNOPSIS
This paper describes a sequence of demonstrations which are used in the Department of Adult and Continuing Education of the University of Durham to help general interest students to gain some insight into the structure of the cathedral and similar buildings. Students do not need more than an everyday appreciation of forces and their lines of action.
The models are made from plaster of Paris and scrap material. They can usually be adapted to give quick answers to questions that arise. The connections with significant features of the cathedral are normally discussed more fully during visits which follow the lecture.

DIMENSIONS
1. Athough dimensions are given below, they are not critical. Care is needed however with the form of the arches and the mating surfaces of the plaster of Paris voussoirs.

FOUNDATIONS
The need for footings.
2. An attempt is made to stand a piece of wooden dowel rod, roughly 1.7cm in diameter and 16cm long, vertically on a block of soft plastic foam. Even though the ends are flat and perpendicular to the long axis of the rod, the attempt usually fails. If volunteers from the class do better, it is with obvious difficulty. The task is much easier if a wider block of wood is interposed between the end of the cylinder and the foam. This usually leads on to the reasons for the difference between the two situations and a discussion of the discoveries at the east end of the cathedral in the 1890s. The stratum on which the building is placed must yield uniformly.
3. Two blocks of plastic foam of the same size and colour but having different elastic moduli are placed side by side and a tower of children's beech bricks built across the crack. Often nothing happens, so the tower is unashamedly loaded with a flat-iron and/or a large can of nails until it demonstrates unequal subsidence and collapses. Light buildings, such as the church of St. Mary-le-Bow survive on inauspicious strata to the east

of the cathedral where the cathedral itself could not.

PILLARS AND ROMANESQUE THICK-WALLED BUILDINGS

4. A sheet of A4 paper is wrapped round a plastic foam cylinder to form a tube about 6cm in diameter and 21cm high. A strip of Sellotape along the edge of the paper prevents it from unwinding. The arrangement can support a 2kg flat-iron comfortably. If the sponge is removed the paper buckles easily. The sponge itself can support nothing. This leads to a comparison of the rubble cored pillars and walls with soft-centred boiled sweets and a discussion of the need for capstones and corbels. It is usually appreciated that the model represents this masonry technique in an extreme way, but that the skins of walls and pillars do carry much of the load.

ARCHES

5. It would be interesting to build arches on wooden centering and try to remove the latter as a demonstration. This would probably lead to an understanding of why so many structures collapsed during this operation. The following technique however, as long as it is made obvious that it was not used in the actual building, has several advantages. It is fast: a series of arches can be assembled before the presentation and the technique can also be used to show how some arches fail.

Setting up arches.

6. A rectangular plywood board, a little larger than the arch to be built, is fixed perpendicular to a substantial flat base, running the length of one long edge. The base should project at least 7cm on one side of the board and the whole is painted black to contrast with the plaster of Paris. The board is supported horizontally and the arch is assembled on it so that when the board with the arch is raised slowly to the vertical plane the arch will be standing on the projecting part of the base. Although the fingers can be used to steady the arch on the way up, stable forms of arch can be left to stand unaided. The base should not be sanded or polished.

The basic arch and its outward thrust.

7. The first model is of a primitive semicircular arch. The voussoirs were made of plaster of Paris cast in plasticine moulds formed on a wooden pattern. Eight of them form a semicircle with internal and external diameters of 13cm and 19cm respectively. The distance from back to front is about 4cm. This is just large enough for a class to see but small enough to pick up, using both hands, without dismantling it.

8. After showing that this arch is stable, it is transferred from the board to a polished surface such as perspex or formica. If it is picked up with the palms of the hands over the top of the arch, then each hand can grasp four voussoirs and the sides of the palms are in a position to restrain the lowest voussoirs on each side from sliding outwards when the hands are opened to release the top of the arch. The open palms can then be moved

slowly apart to show how the arch pushes outward against them. Methods of restraining this outward thrust are demonstrated later. Returning the little arch to a rough surface indicates that friction is restraining the arch here.

9. It is obvious to the class that the arch has an even number of voussoirs. Usually there will be someone surprised that there is no central voussoir and that it stands without a keystone. It might help to insert and hold firmly a thin sheet of aluminium or tinplate vertically between the top two voussoirs and remove one half of the arch. They may then see the arch as two quadrants leaning on each other. Quadrant arches are significant in Durham and there is an unrelated example on the banks to the west of the monastic buildings. An important function for keystones is seen later.

Thin semi-circular arches need help.

10. The next arch is made in the same way as the first but there are 28 voussoirs forming a half circle of approximately 36cm external diameter and 29cm internal diameter. The faces in contact are 3.3cm squares. The board on which this arch is assembled is only slightly wider than the arch so that it is possible to keep the thumbs behind it and the finger tips on the shoulders of the arch to prevent collapse as it is raised to the vertical plane.

11. The class will take the frictional restraint at the springing for granted and can concentrate on what happens when the restraint at the shoulders is slowly relaxed. The top of the arch flattens and shows cracks on the intrados while the shoulders push outwards.

The points that can be made are:-
(a) Restraint is needed at the shoulders of a semicircular arch. In the arcades this is provided by the spandrels. In the vaulting of the nave, the restraint is provided by rubble, filling the pockets in the vault almost to the level of the crown.
(b) *Any* sort of spreading will cause the crown of the arch to flatten and cracks to open on the intrados. An example can be seen in the transverse arches in the south transept.

12. The model arches sometimes cause trouble when the voussoirs rotate on high spots in the joints. This provides an opportunity to explain the role of mortar in taking up imperfections in surfaces in contact. Occasionally someone will question the assumption that mortar should not be considered in the discussions about stability.

What shape of arch *is* stable?

13. The thin semicircular arch is clearly not stable and at this stage simple arguments are used to suggest that the ideal shape for an arch of uniform cross-section is a catenary curve. One such argument, presented with a lot of rapid sketches, starts with the problem of supporting a stone by squeezing it between two others in the manner of the party trick except that the surfaces in contact can be inclined. It then considers what direc-

tion the thrust must have in order to support those three stones between another two, then five between another two and so on. The parallelogram of forces does not have to be mentioned by name. There is enough shared experience of dragging sledges to allow people to be convinced that the line of thrust becomes more nearly vertical as the number of stones to be supported increases. The point seems to be easily accepted that, for stability, the line of thrust must always be parallel to and within the arch as it grows, so to speak, from the top down. The tensions in a hanging chain are clearly parallel to and contained within it and, if we could somehow freeze the joints and invert the chain, so would be the compressions. Replacing links by voussoirs, an arch of this shape should be stable.

14. While classes say they are convinced by these arguments, some form of test is essential. This is done with a model.

A model catenary arch

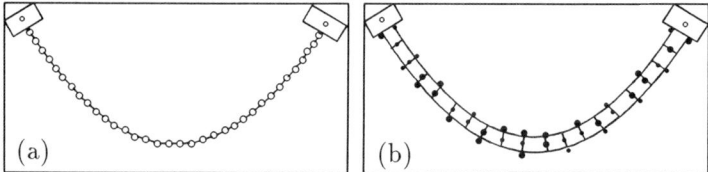

Fig. 1. Making a catenary arch.

15. Two blocks of wood about 15 x 5 x 5cm are mounted with large coach bolts on a board about 1m x 0.5m x 12mm, near opposite ends of a long edge as shown in Fig. 1(a). The shape of a catenary curve is found by standing the board so that the blocks are at the top and hanging a chain, about 1m long between the centres of the lower faces of the two blocks. Since these blocks are to be the abutments of the arch, they are rotated until they are pependicular to the ends of the hanging chain and then fastened down firmly.

16. After marking the curve on the board, the chain is removed, the board is laid flat and a trough 5cm wide and 7.5cm deep is built on it to contain the catenary curve centrally, using thin plastic sheet or cardboard, and plasticine or masking tape to fix and seal it. The trough is then divided into sections of roughly equal length by pieces of the thinnest SRBP material available, held perpendicular to the axis of the trough by plasticine (Fig. 1(b).) The whole trough is filled with plaster of Paris and, after this has set, the segments are numbered before being separated.

17. When the segments of the arch are reassembled, the board is raised so that the arch stands on its abutments. It is found to be very stable and capable of supporting dramatic loads, although not without some anxiety on the part of those watching, after their experience with the semicircular arch.

18. The effect is enhanced if part of the board is made detachable so

CATHEDRAL ENGINEERING

as to leave only the abutments, a strip of wood connecting them and the arch standing clear.

19. In practice the casting of the catenary arch is done long before the presentation and it is only necessary to describe the casting method and to hang up the chain within the outline of the arch marked on the board to show that the arch is indeed a catenary curve.

Why did the small arch stand up?

20. If the inverted catenary is the correct shape for an arch of uniform cross-section, we have to ask why the small arch was apparently stable. A sketch is used to show that its thickness is so great in relation to its diameter that it is possible to draw a catenary curve within it. It is effectively a catenary arch with some excrescences making its outline semicircular.

Arches of other shapes.

21. The argument is extended to chains and arches of varying weight per unit length. We can stabilise an arch of virtually any convex shape we like by loading a chain till the shape is obtained and then ensuring that weight is distributed along our arch in exactly the same way.

22. With help from a volunteer, the original chain is held up with a bunch of keys at its centre point. Something not far off the 'gothic' pointed arch is obtained. Most people then see that a stable arch of this shape needs a large load at the crown and grasp one function of the keystone.

CONTAINING THE THRUSTS OF ARCHES AND VAULTS

23. It is clearly inconvenient to build and rebuild arches to demonstrate their thrusts. Instead, hinged strips of wood are used to produce outward thrusts. Each consists of two pieces of wood about 20cm long with chisel points at the ends and hinged by strips of Sellotape. Three cross-sections are used; 1.2 x 2.1cm, 1.2 x 3.4cm and 1.9 x 4.5cm, so that thrusts can be varied. Care is taken to point out that these are *not* roof trusses, merely a device to give a controlled thrust. The walls to which the thrusts are applied are represented by pairs of wooden blocks of different sizes.

High vaults are more difficult to contain than low vaults.

24. Two blocks of wood, marked **A** in Fig. 2, of 4.4cm square cross-section and 18cm long, have steps cut on one short edge to take the ends of the hinged devices described above. Using the medium sized hinged rod placed as shown in Fig. 2, the blocks are spaced so that they can be relied on to topple. This distance is determined before the presentation and in this case is about 30cm

25. This test is repeated with two blocks, marked **B** in Fig. 2, of the same cross-section and separation as blocks **A**, but only 5cm high. They do not topple. If an explanation is asked for, it can be attempted by indicating how thrusts from the 'vault' in the two cases combine with the weights of the walls. The resultant passes into the ground within the base of the low and stable 'wall' but passes out of the taller 'wall' above the ground where it tends to turn it about its outer lower edge.

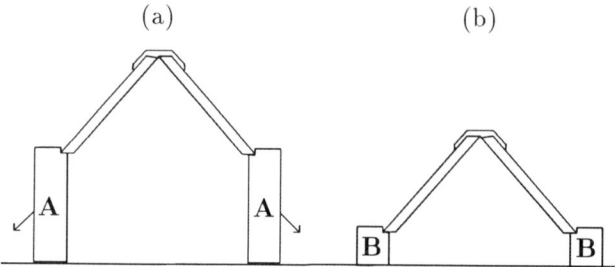

Fig. 2. High and low vaults

How can we contain the thrusts of the vaults?

(a) Use thicker walls. Two lengths of fence post 10 x 10 x 40cm are used as 'walls' and of course cannot be moved by any of the thrusting devices. The fence post sections are also used as raised rough surfaces in the other demonstrations and as walls for a barrel vault.

(b) Use buttresses. The medium hinged strip and blocks **A** are set up as shown in Fig. 3(a), at the same distance as before. Two blocks of wood 4.5 x 7 x 12cm stand against them and the system is stable.

(c) Instead of standing the buttresses against the walls, stand them *on* the walls as in Fig. 3(b). The system is stable and is used to draw attention to the height of wall above the springing of the vaults and its contribution to stability. Reference can also be made to the use of pinnacles and heavy gables.

(d) Flying buttresses. The blocks **B** are placed about 8cm from the blocks **A** as shown in Fig. 3(c). Two pieces of 7mm plywood, 11.4 x 5.6cm are placed on the steps of the smaller blocks and lean, without fastening against the taller ones. The system is again stable. This is used to illustrated several features of flying buttresses although it is not agreed that this would be relevant to Durham.

(e) The use of very light vaulting can be illustrated with the smallest of the hinged strips. With growing expertise and the passage of time, vaulting does become thinner. This appears to be the case, even along the length of the nave in Durham.

(f) 'Turning the walls sideways'. Very thin walls are represented by blocks of wood 2 x 4.5 x 17cm. Although very unstable with the lightest of the hinged boards, these blocks can be turned so that the broad surface is parallel to the line of thrust. For this purpose these blocks have two steps, one on the broad face and the other on the narrow one. This is used as a very crude picture of the structure of later gothic buildings where there are large areas of glass, small areas of solid wall but huge areas of buttress, or of buildings where the dividing walls between lateral chapels act as internal buttresses.

CATHEDRAL ENGINEERING

(g) Tie rods. An aluminium strip, about 2 x 32cm and thick enough not to sag, has 1cm at each end turned up. The ends of the strip rest on the steps in the blocks **A** and receive the thrust from the heaviest of the hinged boards. From a seat in the class this is a satisfactory picture of a tie rod and the aesthetics and desirability of this form of restraint can be considered.

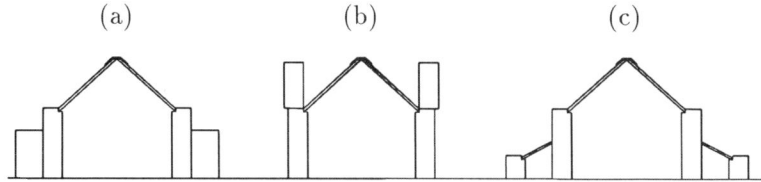

Fig. 3. Buttresses and Pinnacles

VAULTS

26. Models of vaults are being developed. One that has already been used consists of about 100 voussoirs of plaster of Paris, 4.5cm long and of such a cross-section that eleven of them form a semicircular arch of 13.8cm internal diameter and 22.8cm external diameter. The depth to span ratio is such that no extra loading is required.

Simple barrel vault.

27. The two sections of fence post are laid horizontally, parallel to each other with a 14cm gap between them, to represent the space to be vaulted. The first vault is built of parallel arch rings in the manner of some Roman vaults, each ring being assembled in the same way as the first small arch above and carried on the board to its position across the gap. It is possible to lift and slide the arch ring off its ledge holding only the two lowest voussoirs, leaving it freestanding. It is not necessary to continue in this way for more than two or three rings.

Vaults without centering.

28. Building vaults without centering has a perennial fascination and this can be done, with care, with the model. Two rings are set up as described above and alternate courses are displaced horizontally so as to provide keys at both ends. It is a relatively simple matter to add extra rings, one voussoir at a time, making sure that the courses at the springing are somewhat in advance at each stage. Each new ring provides the keys for the next and a long barrel is quickly set up, with an assistant working at the opposite end. Collapse threatens frequently and the tension, and therefore the interest, is occasionally high. The ring of keys could be provided around a window or arch so that the vaulting web could be *started* without special centering.

Holes in vaults.

29. With care, gaps can be made in opposite sides of the barrel vault, starting with a single voussoir at the level of the springing and gradually

widening the holes until two parts of the vault are connected by three voussoirs at the top. The vault and its supports are normally set up on a large board which can be rotated to allow the class a clear view of the apertures produced. It is a short step to a consideration of groined vaulting and pierced barrel vaults. The untidy and difficult form of the groin edge makes the addition of ribs as an aesthetic device seem to be very natural. The Durham vaults went far beyond this however and models are planned to illustrate the use of ribs as a constructional technique.

CONCLUSION AND ACKNOWLEDGEMENTS

People with limited experience of mechanical sciences but an interest in the cathedral say that the sequence of demonstrations makes sense of a lot of things that previously were just words. The dubious stability of some models and the unexpected stability of others appears to provoke considerable interest.

The use of plaster of Paris was suggested to me, more than 15 years ago, by Dr. Michael Rowell's account of a school project on bridges.

I would like to thank Dr. John Senior who initiated the course of which these demonstrations form a part, for the opportunity to participate in it, and both him and Mr. Pat Musset for their lively contributions during the presentations.

We are all grateful to the Dean and Chapter of Durham Cathedral and their officers for allowing us and our students to see the vaulting from the roof space.

The geological setting of Durham Cathedral

G. A. L. JOHNSON, PhD, DSC, FRSA, FGS, Emeritus Reader in Geology, University of Durham

SYNOPSIS. A recent programme of cored boreholes in and adjacent to the River Wear gorge at Durham has proved the bedrock succession in the vicinity of Durham Cathedral. Coal Measures strata, including hard sandstone up to 10m thick, form a good foundation and provide excellent freestone for building. The Cathedral has had an enviable history of stability over the past 900 years. Structural movement is mainly restricted to the extremities, the original east apses, the Galilee Chapel and the medieval North Porch. The main fabric, although in need of constant maintenance, is not giving cause for undue concern today.

INTRODUCTION

1. The Durham Peninsula, a narrow tongue of high ground almost surrounded by the deep gorge of the River Wear, is a natural fortress. Originally, the only access was a steep ascent from the north that led to a rocky plateau about 730m long sloping gently towards the south. The narrow Peninsula is only 240m wide between the two reaches of the river at the north end and here defences such as ditches, banks and palisades could easily be formed. The earliest inhabitants were probably nomadic hunters carrying flaked tools of grey and white flint that have been found on both sides of the river gorge. Later, agricultural settlers established a stronghold on the Peninsula behind northern defences and obtained protection here from Norwegian, Danish and Scottish raiders. The community of St. Cuthbert, a Saxon religious order, arrived in 995 AD, bearing the body of their revered Saint to this place of safety. Already the fame of St. Cuthbert had spread widely and the remains of the Saint, incorrupt after death, was much venerated and treasured. By September 999, sufficient of a new stone church had been built for it to be consecrated and the body of St. Cuthbert enshrined within. This minster, constructed of light coloured freestone, was much admired and won the name of the White Church.

2. After the Norman conquest, William the Conqueror against much opposition brought strong rule to the north. He came to Durham and ordered a great castle to be built on the site of the earlier Saxon defences. The second Norman Bishop, William of St. Carileph or St. Calais determined to replace

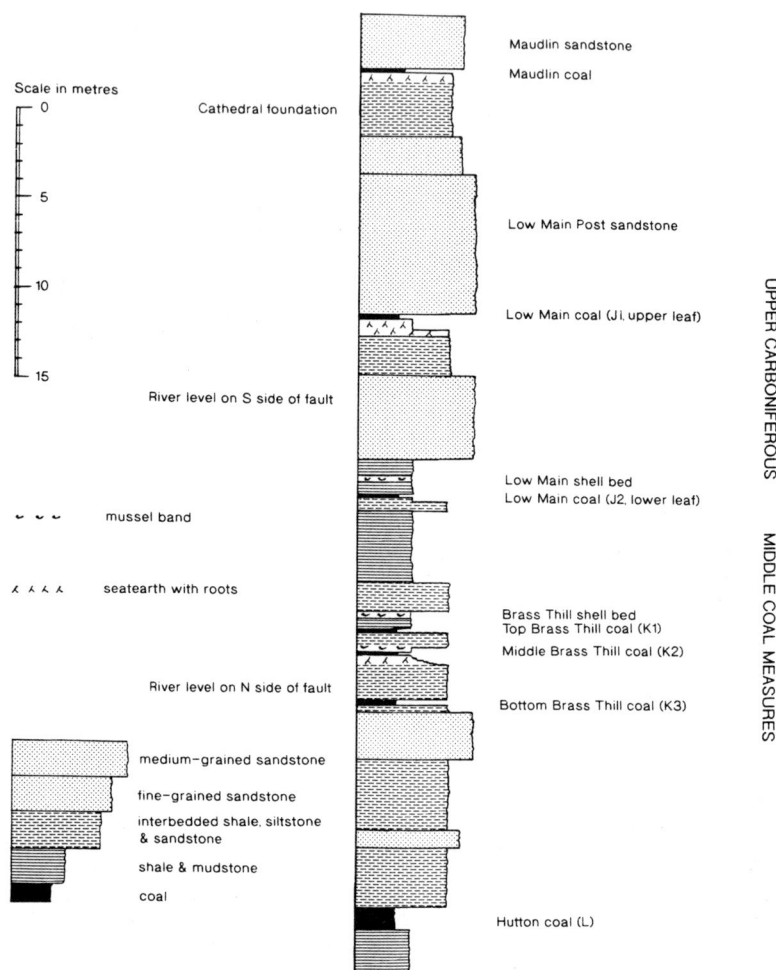

Fig. 1. Coal Measures succession in the River Wear Gorge, Durham City.

GEOLOGICAL SETTING

the White Church with a new and finer minster and brought the Benedictine rule to the Durham monastery. The White Church was taken down in 1092, foundations for the new Cathedral were started in July 1093 and the foundation stone was laid on the 11th August in the same year. The third Norman Bishop, Rannulph Flambard, continued Carileph's building work on the Cathedral. He further protected the site by surrounding the top of the Peninsula, an area of 23.5 hectares (58 acres), with massive stone walls containing case-mates and sentry walks. He also connected the Peninsula with the west bank of the river by erecting Framwellgate Bridge in stone. Some years later Elvet Bridge was built by Bishop Hugh du Puiset.

3. All this work on Durham defences, bridges and large prestigious buildings asserting Norman rule over the region, required building stone. In this the Normans were fortunate, as an excellent freestone 10m thick was exposed in the gorge of the River Wear on the margins of the Peninsula. Probably, it was already well known because it had been quarried by the earlier Saxons for their White Church. This sandstone, called the Low Main Post, seems to have been used almost exclusively during building the original Cathedral fabric. The builders appear to have taken much trouble to locate new sources of Low Main Post as reserves became depleted in the original quarries.

FORMATION OF THE PENINSULA

4. The origin of the celebrated loop in the River Wear at Durham, which almost isolates the Peninsula, lies in changes of river course during the late stages of the Great Ice Age or Quaternary glaciation. The Wear valley and its tributaries were deeply buried in gravel, sand and clay at this time as it is still on the eastern and northern sides of Durham. The late glacial river was in a wide flat valley over which gravel and sand terraces formed and the river took a sinuous route in tight meanders. Probably, when the river was forming terraces at about 53m AOD at St. St. Oswald's Church and under the University Science Laboratories, it deviated out of its ancient course and on to the old valley sides where rockhead was nearer to the surface. Just downstream from the position of Elvet Bridge the river cut through soft surface deposits or drift into hard bedrock. This hard rock persisted downstream round the horseshoe bend to just south of Framwellgate Bridge where deep drift is again present in another buried valley running down the North Road and in front of the Castle. When a river cuts down to bedrock it can no longer migrate laterally and form meanders, but can only cut downwards to smooth its course towards the sea. In the many centuries since the river reached bedrock at Durham, the land has tended to rise and the gorge has gradually been cut by the river round the Peninsula. The loop or incised meander of the River Wear at Durham is the captured path of a late Ice Age river.

BEDROCK SUCCESSION

5. The bedrock succession in the vicinity of Durham is almost everywhere masked by unconsolidated glacial and alluvial drift that reaches more than

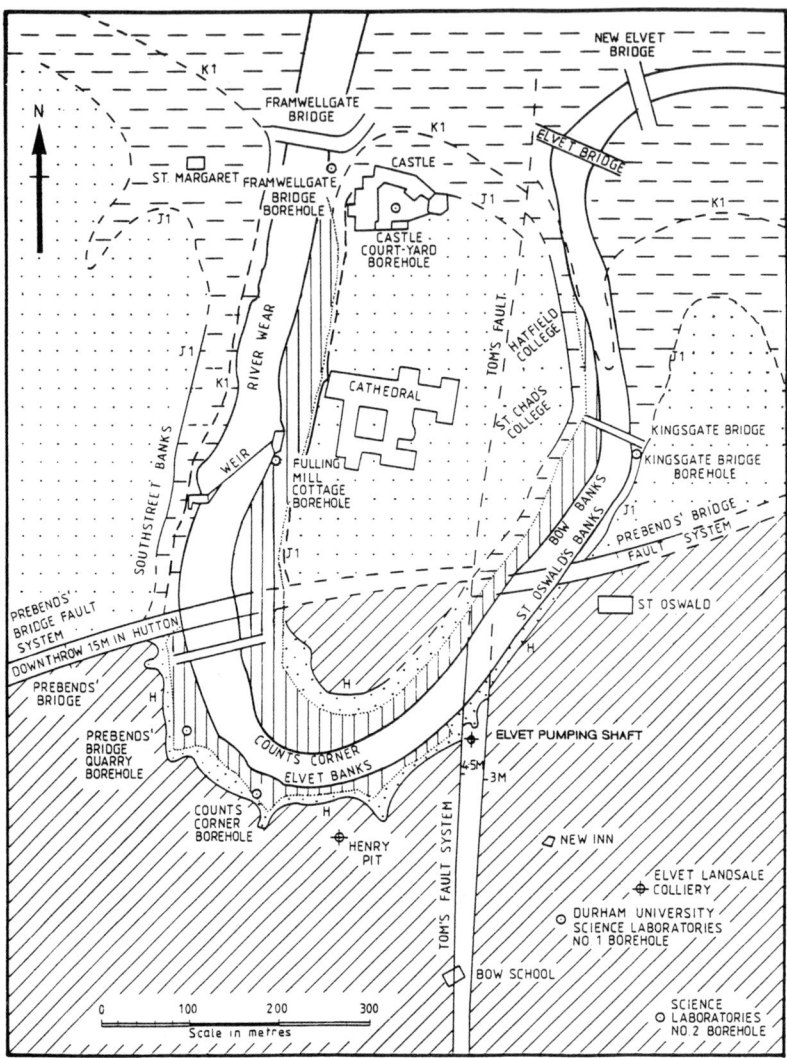

Fig. 2. Geological map of the River Wear Gorge, Durham City. Low Main Post sandstone (stippled), the Maudlin sequence above (oblique lines), the Brass Thill sequence below (horizontal dashes) and ancient quarries in the gorge (vertical lines). Coal seams are identified by their standard index letter as follows: K1, Top Brass Thill; J1, upper leaf of the Low Main; H, Maudlin seam.

GEOLOGICAL SETTING

60m thick in some of the buried valleys. The only natural exposures of the underlying solid rocks occur where the rivers Wear and Browney have cut through the drift in parts of their courses. The gorges on the Wear at Durham and Finchale and the banks of the Browney at Baxter Wood and Burn are particularly noteworthy and are places where the Low Main Post sandstone is exposed and has been quarried.

6. The bedrock succession consists of sandstone, siltstone, shale and coal seams of Upper Carboniferus, Middle Coal Measures age, formed in a delta top environment about 320 million years ago. These rocks were first described from the Wear gorge at Durham by Holmes (ref. 1). He recognised three groups of coal seams in the gorge, from the base - the Brass Thill, Low Main and Maudlin seams all of which tend to be multiple in the Durham area (fig. 1). Two thick sandstone bands are present. The Low Main Post above the Low Main coal is 10m thick and the Maudlin sandstone above the Maudlin coal is 3m thick. Subsequently the region was surveyed by the Geological Survey (ref. 2) and it was found that detailed correlation of the Coal Measures succession in the region was difficult owing to lack of bedrock exposure. To provide the required stratigraphical details, ten diamond drill holes were put down around the Peninsula between 1964 and 1984 (fig. 2) and the logs were publsihed by Johnson (ref. 3). A revision of the geology of the gorge was later produced by Johnson and Richardson (ref. 4).

7. The cause of the stratigraphical uncertainty proved to be the Prebends' Bridge fault that crosses the gorge from just north of Prebends' Bridge to just east of St. Oswald's Church (fig. 2). The fault has a downthrow south of more than 15m and by coincidence this movement brings the Maudlin coal on the south of the fault to the same topographical height as the Low Main coal on the north of the fault causing apparent continuity. Owing to thick cover of colluvium and vegetation, the Prebends' Bridge fault is not clearly visible in the gorge and the juxtaposition of Low Main and Maudlin coals was unsuspected until the drilling programme was completed (ref. 4).

8. Below the Galilee Chapel at the west end of the Cathedral the bedrock succession is partly exposed and the full sequence can be deduced. The Low Main Post sandstone can be seen below the city wall both north and south of the Galilee Chapel and the base of the sandstone with the Low Main coal 0.45m thick is exposed at the foot of the Galilee Well described by Fowler (ref. 5) and Johnson (ref. 6). The steep slope below the sandstone outcrop that falls 30m down to river level is composed of fill banked up against ancient quarry faces. Although concealed beneath a curtain of fill, the sequence below the Low Main coal can be deduced from exposures further north along the side of the gorge and from the opposite (west) bank of the river (fig. 1). The succession of sandstone, siltstone, shale and coal seams is similar to elsewhere around the Peninsula except that the sandstone band between the upper and lower leaves of the Low Main coal thickens northwards to about 4m thick below the Galilee. This sandstone outcrops in the ancient Broken Walls Quarry 100m north of the Galilee Chapel and it can be seen again in a vertical quarried face behind shops at the east end of

Framwellgate Bridge. A useful stratigraphical marker, the Brass Thill Shell bed (fig. 1) has been reported visible in the cellars of the old Castle Hotel, now shops at the east end of Framwellgate Bridge, by Dunham and Hopkins (ref. 7). The shell bed and the associated Brass Thill coal seams have also been seen in temporary excavations on the west side of the river.

DRIFT DEPOSITS

9. Durham city lies in a basin dominated by rounded hills that rise steeply to a plateau at about 100m AOD. This topography is produced by erosion of glacial and glacio-fluvial deposits laid down during the Quaternary Great Ice Age. Geologically this is a very recent episode which only came to an end some 10,000 years ago. The standard sequence of drift deposits in the region has three divisions according to Francis (ref. 8). A lower grey boulder clay covered by the Middle Sands, a thick sequence of interbedded sand, gravel and clay, overlain by a brown coloured stony clay named the Upper Boulder Clay, but it may have been formed at least in part by solifluction. The sequence varies in thickness, in the Durham area, from a few metres over regions where the bedrock is near to the surface to over 60m in the ancient buried valleys. The Lower Boulder Clay dominates the deeper parts of thick drift sequences, the Middle Sands form the many rounded hills which surround Durham City and the upper Boulder Clay forms the plateau surface at 100m AOD.

10. On the Peninsula, both the Lower and the Upper Boulder Clays are absent and a reduced thickness of sand and sandy clay with bands of gravel varying from under 1m to 9m thick are present. In a borehole in the Castle courtyard, drilled in 1904, 8m of sandy drift were recorded (Borings and Sinkings No. 2655, ref. 9) and in preliminary investigations for Kingsgate Bridge, over 9m of mainly sand and sandy clay was seen by Collard (ref. 10). Excavations under Jevons House, Hatfield College, adjacent for the North Bailey, showed made ground and sandy clay 8.2m thick at the north end of the exposure and 4.2m thick at the south end near to the College Chapel, in both cases resting on sandstone. Under Hatfield boiler house, 50m further south and only 60m NW of the Nine Altars Chapel of the Cathedral, 6m of interbedded sand, gravel and clay resting on sandstone were seen by Mockler (ref. 2, p.234). Rockhead appears to be uneven and nearer to the surface away from the northern and eastern margins of the Peninsula.

11. This was confirmed by excavations at the North Door of the Cathedral during the summer of 1992. On the east side of the door, 1.52m of made ground rests on 0.43m of yellow coloured stony clay and wet sand over yellow and white sandstone bedrock. A second excavation 8.8m west on the western side of the North Door showed 1.5m of made ground on 0.48m of yellow sandy clay with stones on sandstone bedrock. About 3m further west further excavations showed rockhead to be rising with only 1.4m of drift over sandstone bedrock. Here, a surface deposit of soil and made ground was 0.4m thick and covered 0.95m of coarse angular sandstone boulders and cobbles in a brown sandy clay matrix containing iron pan layers and coal fragments. Immediately under the eastern side of the north-west tower of the

GEOLOGICAL SETTING

Cathedral, 3.4m further to the west, 0.25m of soil and made ground were seen over 0.21m of angular sandstone clasts in brown sandy clay matrix, but rockhead was not reached. Omitting the made ground that is of recent origin, there is less than one metre of very stony drift above rockhead at the northwest end of the Cathedral. This is to some extent confirmed by the presence of a thin humic soil horizon lying on top of the stony drift, the original soil cover before human activity. It seems likely, that before building activity started, the sandstone bedrock outcropped at the surface in the vicinity of the Cathedral along the western margin of the Peninsula and from this the often quoted description "rocky plateau" came about.

12. On the south side of the Galilee Chapel a case-mate, called the Soldiers' Walk is inclined upwards through the city wall and exposes the Low Main Post sandstone to within a few metres of ground level. The sandstone is also present in the cellars of houses at the west end of the College. This is a southward continuation of the ridge of bedrock at or near to the ground surface on the west side of the Peninsula and drift cover in thin or absent. Drift cover is again absent under St. Cuthbert's Society buildings adjacent to the Watergate on the South Bailey where the Maudlin sandstone and underlying Maudlin coal were exposed during building work. Drift cover seems to augment in thickness eastwards across the south part of the Peninsula and in an excavation in the North Bailey, 32m north of the College gateway, 0.75m of hard core rested on one metre of grey stony clay with sandy clay at the base. From these records of the thickness of drift over bedrock, there seems to be a general thinning of drift from north to south down the Peninsula. The thickest sequences are on the northern and eastern extremities with only thin drift cover over much of the west and south margins.

QUARRYING AND MINING IN THE GORGE

13. Quarrying and mining activity have done much to shape the present day gorge of the River Wear around the Durham Peninsula. Quarrying for building stone began before the Norman conquest and must have increased greatly by the beginning of the 12th Century when the Castle, Cathedral, city walls and bridges over the river were being constructed. The most favoured building stone was the Low Main Post sandstone, a light coloured, well cemented stone that can be masoned and sculptured easily. It was probably first obtained from the east side of the Peninsula, but quarrying spread all round the gorge wherever good stone could be found (fig. 2). The only places where the original steep sides of the gorge remain are places where the sandstone is thin for quarrying, such as behind the Corn Mill, or below St. Oswald's graveyard where the sandstone could not be quarried below church land. When Leland visited Durham early in the 16th Century, he found many quarries on the sides of the gorge and was told that the gorge had been made by quarrying stone and that the original course of the river went on the north side of the Castle; but he states that he does not believe this (ref. 11). There are other historical records of quarries under the city walls and on the other (left) side of the river. Bishop Bainbridge retained the right to quarry stone

when he granted land on the right bank of the river to the Prior and monastery in about 1508. Over a century later building stone, from a quarry in the sandstone between the upper and lower leaves of the Low Main coal, was used for repairs to the Castle by Bishop Cosin in 1663. Obtaining stone from the sides of the gorge may have continued long after this. At about this time, however, the abandoned quarry faces below the city walls had started to be covered with rubbish and fill. Eventually smooth slopes down to the river side were formed and these slopes were cultivated by the householders in the Bailey as gardens. Most of the quarry faces were buried in this way, but the old quarry that worked both the Low Main Post and the Maudlin sandstones adjacent to the SW side of Prebends' Bridge, once known as the Sacrist's Quarry, was still being used as a public rubbish dump in the 1920s.

14. The coal seams that outcrop in the sides of the gorge are split seams that are not sufficiently thick to be mined underground, but elsewhere in the region where the splits coalesce thick seams are found. About 11m below the Bottom Brass Thill coal a persistent seam of high quality bright coal one metre thick, called the Hutton, has been worked throughout the region (Fig. 1). This seam was not mined below the Peninsula for fear that it would cause subsidence and damage the Cathedral, Castle and other associated buildings. However, mine workings extend to the left bank of the river and there are two old mine shafts near the lip of the gorge. Elvet Colliery pumping shaft, SW of St. Oswald's Church, is now marked by a circle of masonry and box drains leading from the shaft can be seen on the river side below. Elvet Landsale Colliery worked the Hutton coal from 1823 to 1908 with the main shafts and offices near the New Inn (fig. 2). Coal was extracted to the south and east of the shafts until about 1870 when a roadway was driven west to take coal from under Nevilles Cross and Crossgate Moor. Henry Pit on Elvet Banks was opened in the early part of the 19th Century and worked the Hutton coal to the south below Elvet Hill and St. Mary's College (fig. 2).

CATHEDRAL FOUNDATIONS

15. A near to bedrock foundation seems to have been found for the Saxon White Church near to the western margin to the Peninsula when building started shortly after 995 AD. The exact site of the church is uncertain, but the later Norman builders of the present Cathedral seem to have chosen the same general position using the ridge of sandstone with only thin or absent drift cover (para. 11). Supporting this, bedrock sandstone is known under the flagstones at the western end of the Cathedral nave. Solid rock rises almost to floor level between the font and the west doorway according to MacKenzie (ref. 12) and this observation was confirmed during the installation of heating ducts in the Cathedral in 1967. Excavations at the North Door and exposures in the Soldiers' Walk both show solid rock under thin drift cover and continue the bedrock ridge to the north and south of the Cathedral (paras. 11 & 12). Hard rock does not persist eastwards below the Cathedral and made ground and sand are found below the floor elsewhere. Drift deposits of made ground, sand, gravel and clay increase in thickness eastwards from under one metre below the north-west tower to 6m in the North Bailey. Much of the

nave, chancel and east transept of the Cathedral are built on drift that increases in thickness eastwards.

16. On the bedrock ridge at the west end of the Cathedral the lowest ashlar courses have been seen at the foot of the east face of the north-west tower. Here the ashlar stone work rests directly on angular broken sandstone in a brown sandy clay matrix at the bottom of the thin drift sequence. The base of the ashlar work is 0.59m below ground surface and about 0.74m above rockhead. The early 12th Century builders seem to have excavated a foundation trench into the bottom drift horizon of broken sandstone, but did not extend their trench to rockhead less than one metre below. It seems unlikely that foundation trenches were taken down to bedrock further to the east where the drift increases in thickness and probably only shallow trenches through the surface soil were excavated. The short period between the start of digging the foundation in July 1093 and the laying of the foundation stone on the 11th August 1093 supports the conclusion that no elaborate foundations were prepared for the Cathedral walls.

17. Despite the apparent lack of foundations, the Cathedral has a reassuring history of stability over the past 900 years since the foundation stone was laid. The main fabric of the church has shown little signs of differential subsidence and it is not giving cause for undue concern today. Some cracking of the nave vault was reported in John Wooler's report to the Chapter in 1777 and this was filled with plaster according to Stranks (ref. 13). It has not given trouble since then. Other defects to the main fabric of the Cathedral, such as bulging walls and collapse of parts of exterior walls have occurred and been repaired on various occasions. These and the failure of the chancel vault in the early part of the 13th Century was owing to structural failure of the building rather than subsidence and foundation problems. It is noteworthy that after the failure of the chancel vault, the new vaulting constructed about 1250 used the original Norman arcading except for alterations at the eastern end; the foundations to the pillars and walls had not failed. Only the extremities of the Cathedral, the original east apses, the Galilee Chapel and the medieval north porch have suffered foundation failure and serious subsidence.

18. <u>The East Apses</u>. Three eastern apses were constructed over the chancel and the two choir aisles during the original construction in 1093 and this was the first part of the Cathedral to be built. By the beginning of the 13th Century the apses and east end of the building were showing signs of collapse and the central apse, behind St. Cuthbert's Shrine, is reported to have been so ruinous that the monks were afraid to enter this part of the building. Instability seems to have affected the whole east end and although it may partly be owing to building failure, foundation problems and subsidence cannot be ruled out. Foundation problems developed on this site when Bishop Hugh du Puiset determined to build a Lady Chapel across the east end of the Cathedral before 1175. Work on the site had made considerable progress when the pillars began to totter and fissures began to appear in the walls. Only shallow foundation trenches had probably been dug, but failure

of the building at this early stage was attributed to St. Cuthbert's preference for male company and the Lady Chapel was moved to the west end of the Cathedral where it, the Galilee, still stands.

19. When Richard of Farnham, architect to Bishop le Poor, designed the east transept or Nine Altars Chapel and started construction 1242, he excavated below the level of the chancel floor and produced the sound foundation that had eluded the previous builders on the site. The floor of the Nine Altars Chapel is 1.2m below the floor of the choir aisles and the foundations for the walls of the chapel go well below this. Possibly the foundation trenches were deepened to below the surface sand and sandy clay into the firmer stony clay and gravel that were reported in the adjacent Hatfield College boiler house section (para. 10).

20. <u>Galilee Chapel</u>. After failing to build his Lady Chapel at the east end of the Cathedral, Bishop Hugh du Puiset moved to the Cathedral yard in front of the Great West Door, well away from the Shrine of St. Cuthbert. Building took place between 1175 and 1177 with the west wall of the chapel being placed on top of the city wall that has a sound foundation on the Low Main Post sandstone. The sandstone rises under the chapel and reaches floor level at the west end of the nave (para. 15), but made ground, at least one metre thick, underlies the Galilee Chapel floor and the position of rockhead here is uncertain.

21. After 250 years the Galilee Chapel was showing signs of movement. The pillars were beginning to lean westwards and there was subsidence. This could well have been caused by prolonged quarrying of sandstone in the cliff below the chapel until the city wall was perched on a narrow ledge, possibly with an overhang below it. Between 1428 and 1435 Bishop Thomas Langley reinforced the crumbling structure by putting on a new roof and supporting the west wall with a series of massive buttresses that are still in place. Langley's work was successful and the chapel remained stable for a further 300 years. When John Wooler gave his report on the state of the Cathedral fabric in 1777 he mentioned that the Galilee Chapel foundations had minor faults, but that they could be put right (ref. 13). On this occasion the remedy seems to have been to stabilise the lower part of the walling and buttresses by burying them in made ground. The base of the walling and the foundations of Langley's buttresses below the Galilee Chapel are no longer accessible.

22. A small amount of subsidence movement can be seen at the north-west corner of the Galilee Chapel at the present time. Again, subsidence westwards towards the river has taken place at some time, but there is no evidence of active movement. It is over 200 years since remedial work was done on the western foundations of the Galilee Chapel and the time is approaching when more work may be necessary here.

23. <u>The North Porch</u>. The medieval North Porch was made necessary by building the Galilee Chapel across the Great West Door by Bishop Hugh du Puiset and so it dates from about the middle of the 1170s. It had a great semi-circular arch surmounted by a pediment in which there were two small rooms above the door. The upper part of this porch pulled away from the Cathedral wall by the 18th Century and John Wooler, in his report to the

Chapter in 1777, recommends that it should be pulled down (ref. 13). This was agreed, the porch was removed and the present doorway was constructed in the 1780's. During excavations around the north door in the summer of 1992, the lowest ashlar course of the medieval porch was uncovered in place. It rests on a thin layer of rough stone hard core and rubble over sandy drift. An inadequate foundation for the medieval porch on wet sandy drift seems to be the reason for subsidence that caused it to pull away at the top, though surprisingly it stood for 600 years.

REFERENCES
1. Holmes, A. The foundations of Durham Castle and the geology of the Wear Gorge. The Durham Univ. Jl., 1928, vol. 25, 319-326.
2. Smith, D.B. and Francis, E.A. Geology of the country between Durham and West Hartlepool. Mem. Geol. Surv. G.B., H.M.S.O., London.
3. Johnson, G.A.L. Records of stratigraphical boreholes drilled in the vicinity of the River Wear Gorge and the University Science Laboratories, Durham City between 1964 and 1984. 1987, Dept. of Geol. Sciences, Durham University.
4. Johnson, G.A.L. and Richardson, G. Coal Measures of the River Wear Gorge at Durham, England. Trans. nat. Hist. Soc. Northumbria. 1990, vol. 55 pt.2, 84-96.
5. Fowler, J.T. The Galilee Well at Durham. Trans. Architectural and Archaeological Soc. Durham and Northumberland. 1907, vol. 5 (for 1896 to 1905), 25-27.
6. Johnson, G.A.L. St. Cuthbert's Well and the Galilee Well. 57th Annual Report of the Friends of Durham Cathedral. 1990, 36-39.
7. Durham, K.C. and Hopkins, W. Geology around the University Towns: Durham area. Geol. Assoc. Guide 15, 1-14.
8. Francis, E.A. Quaternary. Pp. 134-152 in Johnson, G.A.L. and Hickling, G. (Eds), Geology of Durham County. Trans. nat. Hist. Soc. Northumberland, Durham and Newcastle Upon Tyne, 1970, vol. 41, no. 1, 5-152.
9. Anon. An account of strata of Northumberland and Durham as proved by borings and sinkings. North of England Institute of Mining and Mechanical Engineers. 1878 to 1910, 4 volumes.
10. Collard, G. Some foundation problems encountered during the current University building programme. Pp. 113-120 in Dewdney, J.C. (Ed.), Durham County and City with Teesside, Durham: British Association.
11. Page, W. The Victoria History of the Counties of England: Durham, 1928, vol. III (City of Durham).
12. Mackenzie, E. and Ross, M. A view of the County Palatine of Durham. 1834, vol. II, MacKenzie and Dent, Newcastle upon Tyne.
13. Stranks, C.J. This Sumptuous Church. 1973, S.P.C.K. London.

Deterioration of the exterior stone at Durham Cathedral

P. B. ATTEWELL

SYNOPSIS. From the time that the first stone of the present cathedral was laid on the 11th of August 1093 up to the present day, and in spite of repeated stone replacement programmes which allowed most of the outside cathedral stone to be re-dressed or replaced by the year 1860, the walls of the edifice have been subjected to constant contour scaling degradation. Some of the factors which have contributed to this deterioration are reviewed and the results of some laboratory tests on stone that has been removed during the current restoration programme, and on replacement stone, are described. It is concluded from the tests that the durability quality of the stone being removed is generally higher than the quality of the current replacement stone.

INTRODUCTION

1. Durham Cathedral was built to shelter the remains of St. Cuthbert, the much-revered 7th Century Saint. Until the year 875 St. Cuthbert's coffin remained on the island of Lindisfarne off the north-east coast of England. On leaving Lindisfarne because of the Viking raids, its Bishop and the community travelled widely in northern England carrying St. Cuthbert's coffin with them. The community finally moved in 995 to a tight bend in the River Wear at Dunholm where they built a Saxon stone church to house the Saint's remains. The first stone of the new Norman Cathedral was laid on the 11th of August 1093 in the presence of King Malcolm of Scotland. Forty years later the cathedral was completed and vaulted throughout. Part at least of the original main edifice is founded on solid rock of the Carboniferous Westphalian B division. Sandstones, shales and coals of this sequence are exposed on the banks of the River Wear. The Gallilee Chapel at the west end, housing the remains of the Venerable Bede rests on the old city wall and has suffered some quite discernible structural movements.

2. The main building material for the cathedral was the local Low Main Post Sandstone, but according to Johnson and Dunham (ref. 1) there are some uncertainties as to where it was quarried. They have estimated that 68 190 tonnes of stone, or thereabouts, would have been required for construction of the main building (assuming an average outside wall thickness of 2.1m, the towers walls 1.5m thick, allowing for a fenestration factor of 11.65%, and a stone density of 2.5g/cc) with an additional 12 000 tonnes for the monastic buildings. Possible quarrying sites include: north of Prebends Bridge, below South Street and the other side of South Street between Old Caffynite and Poole Houses of Durham School and the former St. Margaret's School; Frankland and Kepier quarries adjacent to Kepier Gorge. The small quarries at Crook Hall and below Prebends Gate (the latter known as the Dell) can be discounted as stone producers for the original cathedral.

MATERIAL CRITERIA FOR THE SELECTION OF BUILDING STONE

3. The following material properties are important in the selection of building stone: unconfined compressive strength, density, porosity, bedding thickness, discontinuity spacing, durability, cuttability, 'soundness'. Sandstones are durable relative to calcareous stones which are prone to chemical attack by carbon dioxide, and clay stones, which may soften, lose adhesion, and degrade when wetted. High compressive strength is not an overwhelming criterion. It is more important that the stone should be able to resist the action of frost and crystallisation of soluble salts on its exposed face(s). Resistance to this form of damage depends largely on the size, extent and nature (for example, tortuosity) of the void space in the stone, usually expressed as the ratio of porosity to water absorption. Stone having a low porosity, or with large pores which allow absorbed water to drain out quickly, tend to have good lasting properties, whereas a stone having a large number of small pores, perhaps occupying the same pore volume, tend to be susceptible to damage. It is also important that this drainage facility is preserved.

STONE DETERIORATION

4. Remedial work on the original masonry of the cathedral has been carried out since the 13th century and more fully recorded since 1777. By the year 1860 most of the outside stone had been re-dressed or replaced. The external walls have therefore been exposed to between 214 and 131 years of environmental pollution and aggression.

During that time much sandstone weathering, such as in the form of contour scaling, has taken place. There are several likely causes of the deterioration.

5. <u>Wetting and drying</u>. Repeated wetting and drying, together with the expansive force of water, can cause fracturing within the pore space of the rock, exploiting the low tensile (relative to the compressive) strength of the material. A temperature increase from 0°C to 60°C causes a 1.5% water expansion and an associated internal pressure of up to $52MN/m^2$. Winkler (ref. 2) has suggested that a diurnal temperature range of 40°C can cause pressures of around $26MN/m^2$ to develop. For a sandstone rock having an unconfined compressive strength of $40-50MN/m^2$, the equivalent tensile strength would be around $5-6MN/m^2$. Winkler has also shown that at temperatures above freezing point, expansion and contraction of water confined within narrow capillaries could develop sufficient pressure to fragment certain types of sandstone within a year of exposure.

6. <u>Crystallisation of salts</u>. The forms of pollution which have an effect on building stone are discussed by Attewell and Taylor (ref. 3) with particular reference to the Durham area. Of those polluting agents the most common cause of stone decay in the UK is the crystallisation (strictly recrystallisation) of salts within the pore space of the rock and frost damage as a result of water ingress. This internal precipitation out of solution is known as 'cryptoflorescence'. Crystallisation of soluble salts on or close to the surface of the stone results in what is known as 'efflorescence', and this will be visible to the naked eye. The main soluble salts, with the pressures that they can exert, are $CaSO_4.nH_2O$ (gypsum: $100MN/m^2$), Na_2SO_4 (anhydrite: $120MN/m^2$), $MgSO_4.nH_2O$ (kieserite: $100MN/m^2$), K_2SO_4 (arcanite) and $NaCl$ (halite: $200MN/m^2$). Sodium, magnesium and potassium sulphates are the main offenders. Their primary action is one of dilation on hydration, the expansion pressure on the rock fabric exceeding the tensile strength of the rock and so causing it to crack. Small amounts of soluble salts distributed over a larger surface may be concentrated by rain water on small areas where they can promote accelerated deterioration. Salt migration occurs during and after each period of rainfall.

7. Two major factors affect the movement of such salts dissolved in water: capillary properties of the stone and the evaporation of moisture. Capillary action depends on the size of the pore space, atmospheric properties, and the temperature gradient across the stone. The rate of

evaporation increases with temperature, decreases with relative humidity, and increases with the velocity of air passing over the stone surface. In a material where the pore water can move to the surface quickly enough to compensate for moisture loss due to evaporation, the surface will remain wetted until the stone is dried out. In circumstances where water is unable to pass through the stone quickly enough, the surface may dry out whilst the body of the stone is still saturated. This will lead to evaporation taking place within the pore space, and the surface miniscus travelling into the body of the stone at a rate equal to the difference in the rate of evaporation and the rate of water movement to this surface.

8. Since evaporation starts at the surface of the stone, water is drawn from the body of the stone towards this point. In so doing, it carries with it the soluble salts. As a result, most of the salts which are recrystallised are precipitated quite close to the surface of the stone, leading to the formation of a case-hardened crust which, because it contains calcite or calcium sulphite, may act to protect the stone. If, however, this crust contains sodium chloride, sodium carbonate or sodium sulphate, the result may be an accelerated weathering rate because these salts act effectively as weathering agents of fresh quartz.

9. Efflorescence is also caused by the use of synthetic materials such as cement, grout and mortar. It often tends to appear during winter months and disappear over the summer. It tends to be found on parts of buildings such as within the zone of rising groundwater, zones below terraces, zones beneath defective gutters and zones adjacent to tight basal stones. Efflorescence can be noted in the cathedral on the clerestory masonry which was renewed under Bonomi in about 1830. During the renewal of the parapet gutters in 1973-74 a fissure was discovered between the outer casing and the backing. The masonry was repaired, but some efflorescence and staining from the stitching and cement grouting still shows on the outer surface of the stones.

10. Chemical attack. Although chemical attack on building stone can lead to the formation of salts, and so contribute to crystallisation damage, chemical attack on its own can be a major source of decay only in special circumstances. In the case, for example, of calcareous sandstones, comprising silica grains bound by calcium carbonate, sulphur dioxide and carbon dioxide dissolved in rain water react with the calcium carbonate leading to the loss of only a very small amount of carbonate but to a large loss of silica grains.

11. <u>Contour scaling</u>. Most of the damage to the cathedral masonry takes the form of contour scaling. Between about 6 and 16 millimetres thickness of sandstone facing has tended to laminate away from the backing despite the stone having been correctly placed with its bedding planes horizontal and so at right angles to the exposed surface. The backing has remained sound. In the older re-dressed masonry there is again extensive erosion, often in the form of honeycombing or vermiculation, with the line of contour scaling being independent of the direction of the bedding.

12. Several possible causes of the scaling have been suggested. Excessive blunt chisel force when dressing the stone may have 'stunned' ('bruised'; induced microcracking) the surface of the stone, promoting eventual separation of the mechanically-worked surface from the backing. 'Hardening' of a new outer surface of the stone due to increasing exposure to air some time after quarrying causes the outer skin to flake off. (A more technically sound explanation is that of stress relaxation on removal from the mass.) Rain-borne soot, tar and sulphurous deposits from atmospheric pollution, and the constant wetting and drying process, build up salts in the outer skin of the stone and change its mechanical behaviour. Bell and Dearman (ref. 4) have also noted the presence of calcium sulphate in the pores of these crusts even when the stone does not possess a calcareous cement, the implication being that the air is the source of the contaminant.

13. Cycles of wetting and drying, freezing and thawing loosens the outer skin and leads to exfoliation. On the cathedral the east and south surfaces of the east front and buttresses have suffered more deterioration than have the north surfaces which do not catch the sun and are therefore less prone to freeze-thaw cycles in winter.

14. <u>Organic growths</u>. There is some uncertainty as to whether or not organic growths are destructive to the outer stone surface, but it is likely that whereas micro-organisms play only a small part in stone decay higher organisms such as lichens and plants can contribute to decay in some circumstances. The metabolic products of lichens are acidic and can cause damage to carbonaceous rocks, but the damage is rarely serious. Urban and industrial areas rid themselves in the past of lichens and mosses quite easily because these plants cannot tolerate soot or sulphate.

15. <u>Animal activity</u>. Animals may affect stone chemically by the secretion of acids or mechanically by boring or digging.

16. _Influence of rain water_. The corrosive power of rain water is directly related to the pollutants it picks up from the atmosphere. Rain water may become very acidic in areas having high sulphuric and carbonic acid. Soot acts as a buffer, and if it is absent the water acidity will increase.

17. The wash-out rate of ingredients from the atmosphere depends on the rain-drop size, the drop speed and the distance between the cloud base and the land surface. Water at temperatures near to freezing dissolves nearly twice as much CO_2 as does water at 25°C. The higher the quantity of dissolved CO_2, and therefore the more carbonic acid generated, the higher is the solubility potential of carbonates. Dissolved CO_2 also accelerates the decomposition of silicate rocks.

18. Moisture is encouraged into the stone by surface tension and capillary action, small pores giving a greater rate of movement which is influenced by the external environment (temperature, humidity, wind), the thermal gradient across the stone, and the properties of the pores. It is drawn towards the colder side of the stone, and so during the day moisture will tend to move into the body of the stone whereas at night it will move out towards the surface.

19. _Influence of sea water_. Sea water has a pH of 8.1 and a total ion concentration of 36 000ppm. Of the latter, 20 000ppm is Cl, 2790ppm is SO_4 and 420ppm is Ca. Since salt spray damage may extend 25 to 55 kilometres inland it could have, and be having, a very slight effect on the weathering of the cathedral stone.

20. _Temperature effects_. Temperature affects the rate of chemical reaction on and below an exposed surface of the stone, and also causes thermal expansion. This expansion takes two forms. First, the individual crystals in the rock have different coefficients of thermal expansion causing them to move both absolutely and differentially, so setting up local sources of stress within the stone. Halite, for example, can expand by some 0.5% over a temperature range of 0°C to 60°C. These stresses promote cracking, which in turn encourages attack by water. Second, thermal expansion across a temperature gradient within the stone is caused by different rates of temperature change in the stone and the air environment outside. The temperature changes more quickly in the surrounding air than in the body of the stone, so creating the temperature gradient. Such effects are not considered to be dominant in the overall weathering process. On the other hand thermal expansion can be a problem when the expansion coefficients of a stone and a mortar are very different. Cracks proportional to the size of the

stone can be formed between the stone and the mortar. A similar effect can occur if the thermal properties of a replacement stone are sufficiently different from those of the original stone surrounding it.

21. Frost effects. Water expands by one-tenth on freezing. Before damage to stone can occur from this source, about 90% of the pore space needs to be filled with liquid. Thus the proneness to frost damage is increased if freezing is preceded by heavy (and driving) rain. Attack is also more severe if the temperature fluctuates around zero degrees, in contrast to continual freezing, and so more readily promotes an expansion-shrinkage cycle with the corresponding cyclic stress effects. The relatively large pore space in most sandstones (greater than about 0.005mm) means that they are comparatively free from such attack, since drainage takes place more readily. Fine-grained rocks having over about 5% absorbed water can be particularly prone to frost action, but they are relatively insensitive and durable below about 1% absorbed water.

22. 'Heat-island' effect. Sixty percent of the winds in Durham blow in a south-west direction. In spring and early summer, north and south-east winds blow.

23. There is evidence in Durham City of the so-called 'heat-island' phenomenon. A 'heat-island' is defined as an zone, usually centred on large urban areas, where the average air temperature is higher than that of the surrounding countryside. Under favourable night-time weather conditions a regular thermal gradient exists from the outskirts of the City towards the market place, producing a nocturnal temperature increase of up to $2^{o}C$ in the city centre. The increasing car population since 1966 will have promoted this effect. An increase in temperature increases the rate of chemical action, in the present case on the cathedral stone.

24. Surface skin formation. A surface skin (or crust) on a sandstone comes about for several reasons; one is the deposition of airborne particles such as atmospheric soot, tar and fly ash and the other is the recrystallisation of soluble salts close to the stone surface, both being mentioned in para. 12 above. Salt precipitation increases the density of a surface skin by as much as 20% in soft, relatively friable stones, by 5-10% in medium-dense stones, and by only 1-5% in dense stones. Calcium sulphate, a product of chemical weathering, is more soluble in water than calcium carbonate which is present in calcareous stones such as the sandstones (new and old) of Durham Cathedral. The presence of calcium sulphate results in the formation of just this type of hard,

DETERIORATION OF EXTERIOR STONE

impermeable surface skin, which leads to blistering and exfoliation of the exposed surface. Such forms of decay are not only unsightly but are also extremely damaging.

25. In a chemically inert stone, such as some sandstones, crustation may be hard but relatively thin, and may be removed by washing with water. If left, it will become thicker and harder, and may require mechanical techniques for its removal, possibly incurring damage to the stone, as noted below. In the case of stones such as calcareous rocks which, being prone to chemical weathering, are not inert, the sulphate, sulphite or nitrate crystals which are formed each time the surface is wetted percolate into the stone and recrystallise with the atmospheric particulate material. This crust does not totally block the pores, but serves to leave the stone in a state of reduced permeability. The rate of growth of a surface skin reduces with increase in its thickness.

CONSERVATION AND REMEDIATION

26. When stone has deteriorated to an unacceptable degree there are four general measures that can be taken to alleviate the situation.

a) Cleaning

27. There is a widespread objection to this method of conservation since the cleaning process may implant fluids, increase air/fluid pore pressures, induce microcracking, and so increase the propensity of the stone surface for further weathering. Where it does take place there should be a facility for graduating the action under the control of an experienced operative, the cleaning process must not produce by-products that could further deteriorate the stone, and the cleaned surface must be, as far as possible, smooth and free from cracks and any other defects that could accelerate weathering.

28. The following methods of cleaning may be used.

29. <u>Water jets/lances</u>. The dynamic pressure of these devices on a rock, typically sedimentary, the surface of which has already been degraded by weathering, may be to cause accelerated degradation in the future.

30. <u>Steam</u>. This is seldom used because cold water has the same effect.

31. <u>Chemicals</u>. Most chemical cleaning agents either contain soluble salts or react with the material being cleaned to form soluble salts so care must be taken not to contaminate the building with soluble salts; acid cleaners, using hydrofluoric acid which leaves no salts and

often used in conjunction with orthophosphoric acid to reduce the risk of iron staining, may be used, in which case the building is initially wetted thoroughly and then the acid works by dissolving the silica for a period of 20 minutes before the solution is washed off.

32. The policy at Durham Cathedral has been not to permit such cleaning.

33. <u>Caustic or alkaline cleaners</u>. These are used to clean limestones.

34. <u>Grit blasting</u>. With dry grit blasting an abrasive grit is blown under air pressure to scour away the dirt. Wet grit blasting involves introducing water to the air/grit stream.

35. <u>Mechanical cleaning</u>. Carborundum-headed tools of different sizes and textures grind and buff the stone. This method is not much used.

36. <u>Poultices</u>. These are often used to remove dirt off small areas of delicate stone. This would not be suitable application for the cathedral.

37. <u>Laser cleaning</u>. Cleaning with this method would not be practical for the cathedral. Each laser pulse cleans only 1cm^2 of stone.

b) <u>Consolidation</u>

38. This is the process, introduced some 165 years ago, whereby loose stone is compressed. However, some of the loose stone may fall off during the process. It is not considered to be a particularly suitable technique.

c) <u>Surface rendering</u>

39. This method involves covering the weathered surface of stone with fragments of sandstone embedded in a layer of mortar which is keyed into sound stone using copper nails. It is an expensive option which has been used in the north of England at Carlisle Cathedral, but if fragments of treated rock later fall off it can produce an unsightly appearance.

d) <u>Replacement</u>

40. This method of remediation involved cutting back stone that has deteriorated and replacing it with new stone. This was the method that was adopted in the 1977 programme and which is being pursued during the current 'round' of restoration. As a general rule the stone should be laid at its natural orientation; that is, in the case of a sedimentary rock

DETERIORATION OF EXTERIOR STONE

the oldest bed should be placed at the bottom with the bedding horizontal.

Preservation

41. The application of such materials as paraffin, linseed oil or tallow to the surface of stone is an ancient practice for stone preservation. Current treatments utilize modern chemical formulations. Before treatment, the surface of the stone must be clean and strong (consolidated). Once treatment has taken place its life expectancy must be assessed and a suitable routine of inspection and re-application implemented. There are several possible treatments which include the following:

42. Microcrystalline wax. This is a tough crystalline wax which may be rubbed into the warm surface of the stone, but the method is expensive. An alternative treatment is to dissolve the wax in a solvent for application to the stone. Although repelling water, and presumably immobilising soluble salts, it affects the appearance of the stone and encourages the adhesion of dirt. Any calcium carbonate in the stone will encourage the formation of soaps which may bleach or stain the stone. (It is noted that wax was applied to some stone on Westminster Abbey but without any measure of success.)

43. Silicones. Silicones dissolved in organic solvent seem to be preferable to the water-soluble types, and they must therefore be applied to a dry surface. Solvent-based silicones can also be applied to stone surfaces previously treated with water repellents. A penetration of about 3mm from a brush, or slightly more from a spray, could be expected with a stone having a medium porosity.

44. Water repellent surfacing. Of possibly the greatest commercial significance is the process of polymerization of monomers within the stone. Typical systems involve vinyl monomers (methyl methacrylate and butyl methacrylate), epoxy resins and alkoxysilanes.

45. Methyl methacrylate is less expensive and more penetrative, but less flexible, than butyl methacrylate. Polymerization is achieved by heating in the presence of a chemical such as benzoyl peroxide. Unfortunately this method, although perhaps suitable for relatively small, detached pieces of stone, is inapplicable for large expanses of stone walling.

46. Epoxy resins generally have too high a viscosity to achieve the necessary pore penetration, but there are epoxide groups, which include 1,2,3,4-diepoxybutane, diglycidyl ether and 1,4-butanediol diglycidyl ether cured with 1:8 diamino-p-methane, that have low viscosities. Dilution with tetraethoxysilane or tetramethoxysilane serves to reduce further the viscosity. Unfortunately, because of general discolouration, and of reaction of the hardener with carbon dioxide which promotes a white efflorescence, this method cannot be recommended. In areas that are less visibly obtrusive, epoxy resin can be used to fill fine cracks in the stone after cleaning and expulsion of any water from the surface zone, injection being by syringe (fine cracks) or tamping (wide cracks) before the material cures. Thermoplastic resins seem to be generally more desirable as surface protective agents, since some types age quite well without discolouration and remain soluble.

47. Chemicals such as alkoxysilanes are able to penetrate the surface of porous stone material and polymerize, so consolidating the underlying friable material. Such preservatives aim to prevent the access of water and/or render the stone material immune from the action of gases and acids in the atmosphere. The requirements of a satisfactory treatment should be:
    - to prevent water entering, yet at the same time allow any moisture already in the stone to escape;
    - to achieve as deep penetration as is possible;
    - to use a chemical which does not create by-products which could react with the constituents of the stone;
    - to use a chemical the presence of which in no way alters the natural appearance of the stone;
    - to use a chemical which develops a lining around the grains of the stone but without clogging the pores of the stone;
    - to use a chemical that does not create thermal movements in the thin surface layer that are sufficiently different from those of the underlying stone to generate shear stresses which could contribute to failure.

48. Without such careful choice of chemical, surface treatment has been found to have an accelerating effect on stone decay. Penetration in such cases may be only 2 to 3mm, leaving sufficient permeability for moisture, and driving rain in particular, to enter the stone and dissolve salts close to the surface of the stone. With a change in atmospheric conditions, the stone dries out and the salts

recrystallise behind the treated layer. Pressure exerted by these salts and by the water trying to escape from the interstices of the stone, but being prevented from escaping by the less permeable skin, results in the skin becoming weakened and eventually falling away. In addition to these pressures acting against the skin, in times of frost there is the added danger that water which has built up behind the skin will try to escape, freeze, expand, and so exert further pressure behind the surface. A large depth of chemical penetration is important because not only will this reduce the stone's susceptibility to frost action but also because moisture content fluctuates far less at depth in the stone. There is less transportation and recrystallisation of soluble salts.

49. For adequate penetration to be achieved, the chemical treatment should be of very low viscosity and have a low surface tension. Some alkoxysilanes have very low viscosities (similar to that of water) and are able to polymerize at outdoor temperatures within the body of the stone before solvent evaporation returns the solution to the surface of the stone. The alkoxysilanes react with water to form a gel which coats the pore space surfaces and hardens. Hydrolysis precedes polymerization, the water of hydrolysis arising from the body of the stone, or by addition before application in which case hydrochloric acid (1 part) needs to be added to the water (10 000 parts). Curing is by alcohol evaporation over a period of several days, depending upon the ambient temperature. Alkoxysilanes provide a colourless product, but conditions of strong ultraviolet light can lead to their breakdown into a white powder. They also have the advantage of posing no health hazard. A greater depth of penetration - 50mm or more - may be achieved by applying the treatment to dry stone, but this ideally requires several days of hot dry weather which can never be guaranteed in the English climate. In addition to the porosity and moisture content of the stone, the depth of penetration is also a function of the amount of salt present.

50. Some preservatives rely on a suitable solvent which leaves behind a solid residue on evaporation. As in the case of soluble salt recrystallisation, the drying process will cause much of the dissolved substance to be drawn back towards the surface of the stone, so preventing any further deep penetration.

51. A grave disadvantage to the use of such materials is the cost should large areas of stone be considered for treatment. Typically, 4 to 5 litres would be needed to treat stone having an average porosity

of about 20% to a depth of 25mm. When labour costs are added, the operation clearly becomes uneconomic except for the preservation of critical areas of carved stone.

52. In order to overcome the problem of allowing the stone to 'breathe' while keeping water out, the Building Research Establishment in England has developed a treatment called "Breathane" specially for stone preservation. It is a colourless liquid having the viscosity of water, and its main component is trialkoxyalkylsilene. Applications by brush or spray continue until no more of the penetrant can be absorbed by the stone, the actual penetration being about 25mm. Breathane does not form an impermeable coating on the surface of the stone. As the solution sets in the form of a gel within the pore space, an alcohol (methanol) is released and evaporates. On its release, the gas creates voids within the stone and it is these voids which allow the stone to breathe. Short-term tests have shown that it can approximately halve the rate of stone decay (ref. 5). There are, however, several limitations to its use:

- It is expensive, and, as is the case with alkoxysilanes generally, it is unlikely to be used for large areas of stonework[1].
- It is volatile and highly toxic (requiring the operatives to wear face masks and rubber gloves for protection) and so requires professional application.
- Application is an irreversible one (removal is not a practical proposition).
- It does not perform entirely satisfactorily in stone containing large concentrations of common salt (sodium chloride).
- Application is a slow and difficult process; a maximum area of 1m^2 should be treated at any one time and the surface should be dry and thoroughly clean, with any large gaps infilled.

53. There are also other commercially-available stone preservatives for which it is claimed the benefits are similar to those of Breathane. One of these, which has been assessed at Durham University for its

---

[1] Given the porosity of a stone to be n (%volume), and the area A (m^2) is to be treated with a chemical preservative, then for a penetration depth d (mm) the volume V (litres) of preservative that needs to be applied is
$$V = d \, A \, n/100.$$
For the Dunhouse stone used as replacement at Durham Cathedral, n = 18.7%, A = (say) 1m^2, and a desirable depth of penetration d is, say, 50mm, the volume of treatment then required is 9.35 litres/m^2.

waterproofing qualities on Cathedral sandstone and which is readily-available at hardwear stores, is "Thompsons Water Seal". It is a combination of hydrocarbon resins and polyoxyaluminium distearate in aliphatic hydrocarbon (mineral) spirit (in other words, it is a combination of hydrophobic resins in an aliphatic hydrocarbon solvent). Waterseal, like Breathane, has a very low viscosity, so promoting deep penetration, and the manufacturers claim that it "attaches itself to walls of pores and does not stop the stone breathing". Experiments were carried out in the University to verify this important property. One 'drawback' of the material is that it again relies on the evaporation of solvent, depositing solid resins but which, it is claimed, never dry out and remain flexible.

Important repair programmes since 1777

54. 1777 to 1827 Repairs. A major restoration to Durham Cathedral began in 1777 under the supervision of James Wyatt and continued until 1803. Wyatt did not come to Durham until 1794! He presented his proposals in 1795 and was no longer involved in the work after 1798.

55. During the period from 1777 the surveying of the cathedral was undertaken by John Wooler who made the following recommendations to the Great Chapter on 20 November 1777.

" The third great defect I now take the liberty to mention is obvious to everybody and that is the almost universal decay or wasting condition of the stones on the outside of the whole structure. To prevent the wet entering and lodging in the walls and thereby bringing on a more speedy dissolution, and to afford all the remedy that can properly be applied on this occasion, it will be necessary to chip and pare off their outsides to a depth of 1, 2 or 3 inches, as may be particularly required, to bring the upright of the wall to a tolerable even or straight surface at the same time taking out and replacing such stones as are almost totally perished and moulder'd away and filling up the joints and beds of the whole with proper mortar stuck in with chips or splinters of flints and gallets as full as it well can be. It would be proper also while the scaffolding for this purpose is up, to fix on the walls the proper lead wall pipes, to convey the main water from the various parts of the roof to the ground. The walls will thus be brought to as perfect a state of repair as they well can be, and may without any very considerable expense, resist the ravages of time perhaps for centuries to come! I will also mention the necessity there will be at the same time to renew nunmous and side janmbs of a great number of windows, which are also much mouldered and decayed as to be scarcely sufficient to retain a hold in the glass."

56. This recommendation by Wooler resulted in the destruction of the north porch, the chiselling away and re-surfacing of the masonry of the east, north and west fronts of the church, the reconstruction of the upper parts of the northern turrets of the Nine alters Chapel and North Trancept, and the addition of pinnacles to the towers.

57. Detailed examination of the masonry of the nave and choir aisles and clerestories proves that the 1,2 or 3 inches mentioned by Wooler were pared off the surface of the Norman masonry. The average removal seems to have been about 2 inches on the north side of the church, judging by the surviving tide-mark line that represents the old graveyard levels built up against base courses of the Norman work. The chiselling process was carried out only down to the ground level, preserving in the lower courses of masonry an indication of the original dimensions of the building and showing that the corner 3/4 shafts were larger than they now appear. More serious was the effect on the Early English east front of the Nine Altars Chapel where the narrower but deeply-projecting intermediate buttresses did have 4 inches taken off, so reducing their widths from 4 feet 4 inches to 3 feet 8 inches.

58. During this time (1777 to 1803) most of the restoration took place on the east front under the supervision of James Wyatt.. He was also responsible for the renewal of the great east window which was restored to its present form in 1795.

59. This external restoration campaign was started under Wooler and Nicholson and continued by Wyatt and Morpeth. William Morpeth was appointed 'collage Architect' in November 1793, and that he was appointed Architect by the Chapter rather than Clerk of Works has some significance. He soon established himself as a dominant character. During these repairs the high north porch was replaced by the present 18th century Gothic casing to the Norman portal. Modern scholars have doubted that any original 12th century carving on decorative masonry survives externally. The stone of the main portal is brown, but the 18th century outer casing of the porch is a greyer stone, the outer-most chevroned border being in the same stone.

60. William Atkinson succeeded James Wyatt to be architect to the Board of Ordnance from 1813 to 1829. By 1804 there was a change in

attitude towards restoration methods of chiselling which had been used during the Wooler-Nicholson repair era. These were no longer to be used. During his report to the Dean and Chapter in 1804, William Atkinson outlined his plans for restoration.

> "Where stone is very much decayed and to a considerable depth, it should be replaced with new, but in general the mutilated part must be brought out to a proper surface with Parker cement. The old stones should be carefully wet when the cement is laid on, and holes may be drilled to give a stronger tie.
>
> "The cement is particularly useful in restoring ornaments and mouldings and at a considerable less expense than cutting them in stone. It is nearly the colour of the moss upon the Tower. But what is of still greater importance, its brown tint adds highly to the sublimity of the building."

61. Cement[2] was used on the Tower in 1806, but by 1808 the Chapter were becoming unhappy with the results, and the work was stopped. Atkinson ceased work on the cathedral in 1809. The cement was removed in the 1850's.

62. Some of Parker's cement statues with brick cores still survive in the Cathedral Mason's store, and the cement is indeed harder than natural stones.

63. <u>1827 to 1860 Repairs</u>. The next phase of restoration work started in 1827 when Bishop Jenkinson was Dean (1827-1840) and entailed the re-facing in particular of the whole south front of the cathedral in new stone, in contrast to the dressing back of the north front masonry under the previous authorities. Work began on the south front and gable of the Nine Altars and the south west buttress and turret of the Chapel, picking up at the point where Morpeth had rebuilt the upper part of the south east turret in 1812-1813.

64. Work was continued under Ignatious Bonomi (1787-1870). He restored the south end of the Nine Altars, the restoration being carried out in new natural sandstone as had been Wyatt's upper east front.

65. Bonomi said the following to the Dean and Chapter of the Cathedral:

---

[2] Atkinson actually manufactured Parker cement himself at his works in London.

"I don't perceive that any of the old facing stone has been saved by the operation, in fact if it becomes necessary to pare down above three inches in depth (in consequence of the erosion surface) the joints at that depth are not found to hold good and present the appearance of very imperfect masonry. I don't immediately advise the resurfacing without paring but I suggest the property of a trial being made by an experienced mason, in order to ascertain if it be expedient or not."

66. Bonomi appears to have finished work on the cathedral in 1835. Work then began again in the 1840's under the supervision of Robert W. Billings. During this time, restoration on the south side was carried out.

67. In 1842 the tracery was restored. A hard grey stone from Gateshead Fell was used to renew the Triforium window surrounds.

68. In 1849 Pickering reported on the south aisle of the nave:

"According to your request I beg to submit a statement of circumstances which led to the removal of all the old ashlar facing of that part of the south front of the cathedral now undergoing restoration.

"In the greater proportion of the front, that is from the top to about the heads of the lower windows, the decay of the mortar and the outer facing of the stone has completely detached the latter from the solid bulk of the walls. This facing was so loose that it was with very much difficulty great quantities of it could be prevented from falling down upon the cloister roof.

"The eastern part of the remaining portion had not suffered so much from the decay of the mortar; but the ashlars were so narrow upon the bed, and consequently had so little hold of the bulk of the wall behind them, that the support which they gave to the body of the wall was trifling - besides not less than four-fifths of them were standing up edgewise, that is, they were not resting upon their natural beds on which they lie in the earth..."

69. During the 1850's re-facing and restoration of the belfry stage and bell ringers walk were undertaken.

"The upper parapet is higher and bolder than that of cement which it replaces. The buttresses, which had been pared away, have been thickened. The entire upper stage has been

DETERIORATION OF EXTERIOR STONE

faced with Prudham and Dunhouse stone, thicker than the cement that was removed, so that the size of the tower is increased, while the stones extend from 10-15" into the wall and are bonded firmly to the old work."

70. <u>Summary of the Repairs undertaken between 1777 and 1860</u>. During this period, the whole of the external masonry was restored, re-faced and re-dressed. Most of the east front, north side and west end was re-dressed to expose a new surface, much of which needs to be re-pointed and renewed at the present time. The upper part of the south side was re-faced in new stone, but some of this has already deteriorated and suffered surface scaling and is in need of pointing and repair.

71. All the repairs to the cathedral have resulted in a variety of stones, textures and different-aged masonry being exposed together. For example, consider the composition of the east front of the Nine Altars where there are several types of stonework:
- the original 13th century masonry, some of it at the very base of the wall where for much of its life it has been protected by built-up ground levels
- the original 13th century masonry, dressed back in the 1780's, all of it weathered again and some of it now severely decayed and eroded
- new stones inserted as repairs during the 1780's campaign, in reasonable condition, although the iron cramps and bowels of the nook-shaft have caused deterioration in these
- the Wyatt-Morpeth re-facing of the upper stages, buttress caps and Rose Window in which, because of the manner of tooling, but mainly due to the particular nature of the stone, much of the surface is suffering from contour scaling, that is, the outer skin is peeling off although the stone backing remains sound and correctly bedded
- limited areas of renewals and repairs made during the 19th and 20th centuries.

Current Repairs

72. The most recent survey was carried out in 1976 by the present architect. No cleaning of the stone has been carried out, but the following work was specified:

73. <u>Urgent work</u>
- Investigation of a minor pocket of fungus above the south side of the choir vault.

74. <u>Essential work to be carried out as soon as possible</u>
- continuation of masonry repairs to the east front, including work on the Rose Window and the gable over it
- minor repairs to parapet gutters and rainwater pipes (extensive damage has been caused to the stone by poorly-maintained lead water pipes and their down pipes on many parts of the cathedral)
- investigation and repair of part of the east of the north west tower above the Triforium roof
- various other masonry repairs.

General State of the Cathedral Stone

75. The medieval stone facings to the east, north and west fronts, the weathered surfaces of which were so drastically dressed back between the years 1777 and 1795, have eroded seriously once more and have been in need of further extensive treatment, renewals and re-pointing.

76. Even on the upper parts of the east front, and on the south fronts, where medieval stonework was re-faced in new masonry between the years 1795 and 1850, there has been considerable weathering and surface deterioration which necessitates general repair and re-pointing.

LABORATORY TESTS ON CATHEDRAL SANDSTONE SAMPLES

77. Slake durability, acid immersion, permeability, porosity, salt crystallisation and water absorption tests were performed on pieces of sandstone (samples 1 to 5 inclusive) retrieved from the cathedral masonry and on pieces of replacement stone (samples 6a,6b and 6c). These most recent Durham University test results supplement test results obtained earlier at the University, and allow comparisons to be drawn concerning the relative durabilities of the weathered and replacement stone.

78. <u>Sample 1</u>, taken from the south side of the cathedral about 1m above the ground, was representative of the oldest sandstone, dating from the 15th century, and until the 19th century had been protected by an outbuilding. It was a coarse-grained sandstone comprising dark beds

of concentrated iron. Usually, the more coarse-textured a stone the more durable will it be (ref. 6). Ferruginous sandstones tend to weather well (ref. 7 *see also* ref. 8).

79. Sample 2, a medium-grained sandstone with a high iron content and some mica, was taken from the south west tower. A zone about 7mm deep was darker than the rest of the stone and showed signs of some flaking having taken place. Soot had been deposited from the more sheltered part of the pillar from which it had been retrieved.

80. Sample 3, a fine-grained siliceous grey sandstone from Gateshead Fell was obtained from the east side of the cathedral. The stone, which had been difficult to work, was placed *in situ* under the directions of Robert W. Billings in 1842 and was used to restore the tracery and also the Triforium window surrounds.

81. Sample 4, a medium-grained stone, light brown to orange in colour with a high iron content, was for the most part covered in soot. The sample was sculptured and highly-fractured, the fractures not following any bedding trends in the stone.

82. Sample 5 was a well weathered, fine-grained sandstone.

83. Samples 6a, 6b and 6c were all replacement stones from Dunhouse quarry near Staindrop, County Durham. The stone is fine-grained and pale yellow when fresh, and has a mainly silica cement. Iron is present to cause staining under weathering conditions. About 10% of the rock comprises orthoclase feldspar, with most of the remainder being biotite mica.

84. Within the sample, which spanned from the weathered surface to the unweathered interior of the stone, a quite distinct line was visible between the protected and unprotected zones. The latter zone was defined by a 1 to 2 cm thick black line of sooty appearance, the line being visible only on the underside of the stone (the most protected area). The stone showed pitting and an almost circular pattern to the erosion on the surface. Slightly more erosion had taken place along the darker iron-rich beds.

85. The slake durability test (ref. 9) examines the resistance of a sandstone to a number of wetting and drying cycles (under mild abrasive conditions), and so attempts to emulate the natural conditions of wetting and drying to which the rock has, and will be, subjected *in situ*. Each test was conducted over six cycles. After the six slaking cycles it appeared that the replacement Dunhouse samples were the most durable. However they did begin to lose weight significantly after the fifth cycle. Had further slaking cycles been undertaken, and the

samples followed the erosional trends established between the fifth and sixth cycles, then there is evidence to suggest that the original sandstone of the cathedral could be more durable.

86. The relative susceptibility of a stone to acid attack can be assessed by immersing samples in acid for a specific time and observing any degree of deterioration. The <u>acid immersion test</u> that was adopted involved immersing rectangular slabs of the stone, 50mm x 50mm x 10mm, in sulphuric acid having a density of 1.306 Mg/m^3 for a period of 10 days (ref. 7). From both the present series of tests and earlier experiments (ref. 3) the consensus of evidence suggests that while the replacement sandstone disintegrated, the cathedral stone remained sensibly intact, with no residue other than a few quartz grains.

87. There seems to be a direct relation between the <u>permeability</u> and <u>porosity</u> of a stone and the water absorption capacity on the one hand, and the degree of disruption by the action of water-soluble salts on the other. Permeability, which is related to pore size, controls the rate of absorption. However, according to the Building Research Establishment (ref. 7) porosity offers no guide to stone durability because it gives no indication of the way in which the pore space is distributed within the stone; for example, whether there are many fine pores or a small number of coarse pores. There is an inverse relation between the saturation coefficient and the susceptibility of a stone to this form of weathering.

88. Porosity was measured by the Kobe method. The replacement Dunhouse sandstone, although of average porosity at 10.5%, was considerably higher than the porosity of the cathedral stone. Permeability, measured by Ruska gas permeameter, involved forcing a gas of known viscosity through a rock core of known cross-sectional area and length. Whilst an average permeability of 84.2 millidarcys (8.15×10^{-7} m/s) was recorded for the replacement stone the cathedral sandstone was found to have negligible permeability perpendicular to the bedding and only a slight permeability (1.31 millidarcys, equivalent to 1.27×10^{-8} m/s) parallel to the bedding. From the results of a limited series of tests it would thus seem that under conditions of driving rain the replacement sandstone would very quickly absorb its full capacity of water which, under sub-zero temperatures, could promote spalling of the rock surface. Some benefit could arise, however, from potentially-damaging salts being more readily flushed out of these rocks under such circumstances.

89. Salt crystallisation in the pore space of rock is one of the most common causes of decay of British building stones. An assessment procedure is outlined in the American Society for Testing and Materials (ref. 10). The propensity for disruption of a rock fabric is measured by subjecting samples of rock to repeated immersion in a solution of salt and then drying the rock in order to promote crystallisation. Salt composition determines the severity of the crystallisation effect, with sodium sulphate creating the maximum disruptive force during subflorescence (ref. 11).

90. Disturbance of the rock fabric during specimen preparation undoubtedly increases porosity along the rock faces, and especially at the edges of a block sample. Expansion of microfractures will accelerate salt solution uptake, so creating zones of preferential decay. However, building stone will experience similar disturbance during preparation and would be expected to respond in a similar manner to weathering processe, albeit at a slower rate.

91. During early cycles of immersion in the salt solution the total sample weight increased significantly as the rapid intake of saline solution precipitated salt in the pore space. After the fourth cycle, block samples of the rocks showed signs of disruption, the corners and edges becoming increasingly rounded. The faces of the cube samples became pitted and rough, often with opposite faces showing more decay than an adjacent side. Sides containing the greatest density of pitting lay sub-parallel to the bedding planes as defined by prefentially-orientated basal planes of mica crystals.

92. The greater the sample weight loss, the greater is the risk of efflorescence. A general conclusion from this suite of tests is that the replacement stone is more susceptible than the cathedral stone to weathering . The siliceous sandstone, sample 3, was the most resistant.

93. The water absorption test is really an expression of specimen porosity. The greater the amount of water taken up by a stone specimen the more likely it is to succumb to salt crystallisation weathering and the actions of freeze-thaw cycles. The replacement sandstone generally showed higher rates of absorption and higher terminal absorption than did the cathedral sandstone.

CONCLUSIONS

94. Much of the industry in the Durham region is to the east of the city in the coastal areas and these generate significant pollution, the major belt of which runs east north east from Billingham. Although

south west winds tend to prevail, it would be expected that rather more stone deterioration would occur on the east-facing sides of the cathedral. This is confirmed by the amount of restoration work performed on this face. Even the upper parts of the east and south front, the latter also experiencing air pollution, have weathered considerably since their re-facing between the years 1795 and 1850.

95. Easterly winds tend to have lower temperatures. The lower the temperature the more carbonic acid can the air moisture absorb and therefore the greater the potential for weathering. The easterly winds often occur under anticyclonic conditions when the presence of a temperature inversion is likely to limit the upward dispersal of smoke particles. The eastern and southern sides of the cathedral may be weathered further by the action of freeze-thaw cycles during winter, and although experiencing lower temperatures the north-side stone is less exposed to cyclic temperatures. This is also the reason why the north-side stone is less weathered than the west-side stone.

96. Much of the damage to the face of the cathedral has been caused by bad guttering. Restoration of the gutters, thereby reducing the amount of water on the cathedral face, serves to lower the susceptibility of the stone for erosion.

97. Present repair work being carried out to the exterior of the cathedral makes use of Dunhouse stone from Staindrop, County Durham. More recent laboratory work conducted at the University of Durham confirms earlier work to indicate that the stone being removed from the cathedral is of more durable quality than the stone that replaces it. With the exception of the fine-grained silicious sandstone, all the weathered masonry removed from the cathedral was either medium or coarse grained. However, as a quite general rule it would seem from experimental work that the more coarsely textured the stone, the more durable it is likely to be.

98. The best sandstone for the purpose of resisting the elements would appear to have been the fine-grained, light grey silicious sandstone which was used by R.W. Billings in 1849 during his restorations. However, because the general appearance of the stone is very unlike the orange-brown sandstone originally used for construction, its further use as a replacement stone would be tend to be discouraged.

99. A facing stone having the optimum qualities of durability would experience little loss of weight during the laboratory slaking cycles. It would show no effect from being immersed in sulphuric acid, implying

that the stone would be resistant to acid rain. Demonstration of a low porosity and permeability would suggest reduced action from soluble salts in the atmosphere.

100. The engineering options for treating a weathering problem on cathedral facing stone may be summarized as follows:

- Replacement if the facing stone is badly corroded. Choice of stone will be determined by a balance between several factors, such as

    quarry price, accessibility and haulage costs;
    quality, in the sense of durability (weatherability), as determined by appropriate laboratory tests, and appearance (reasonable colour and textural match with unreplaced stone in close proximity).

    In practice, because of the expense involved, replacement at any one time may be limited to relatively small, architecturally and visibly important areas of the edifice.

- Preservation treatment of existing stone insufficiently weathered to require replacement or (less likely) of new stone. Before treatment, an existing stone surfacing needs to be cleaned. In descending order of attraction, the options seem to be:

    water-repellent chemicals such as Breathane (Building Research Establishment) or Thompsons Water Seal (commercially available) which, because of the expense involved, would be used only for specialist areas, such as those incorporating stone carvings;
    thermoplastic resins, silicones, microcrystalline wax, epoxy resins (not recommended).

- Abrading and cleaning of existing stone insufficiently weathered to require replacement and serving two possible purposes: preparation for preservation treatment or the enhancement of the visual aspect of the building. In descending order of merit the options include:

    gritblasting with compressed air;
    mechanical grinding;
    poultices or lasers (for removing dirt from delicate areas of stone);
    chemical cleaners, grit blasting with water, water lances/jets and steam (not recommended for adoption without considerable investigation as to their immediate and long-term effects).

ACKNOWLEDGEMENTS

101. In producing this paper the author has drawn on the project work performed by Susan Chadwick as part of her MSc Advanced Course in Engineering Geology and on the project work performed by Andrew Dugdale as part of his final year undergraduate degree in Engineering, both at the University of Durham under the supervision of the author. Dr G.A.L. Johnson of the Department of Geological Sciences, University of Durham has been kind enough to comment on and correct some statements in the original text of the paper.

REFERENCES

1. Johnson, G.A.L. and Dunham, K.G. The stones of Durham Cathedral: A preliminary note, Trans. Architectural and Archaeological Society of Durham and Northumberland, New Series 6, 1982, 53-56.
2. Winkler, E.M. Stone: Properties, durability in man's environment, Springer-Verlag, New York, 1973.
3. Attewell, P.B. and Taylor, D. Time-dependent atmospheric degradation of building stone in a polluting environment, Proc. Int. Symp. on The Engineering Geology of Ancient Works, Monuments and Historical Sites: Preservation and Protection, 16-23, September 1988, Athens, P.G. Marinos and G. Koukis (eds), A.A. Balkema, Rotterdam, 1988, 739-753.
4. Bell, F.G. and Dearman, W.R. Assessment of the durability of sandstones with illustrations from some buildings in the North of England, Proc. Int. Symp. on The Engineering Geology of Ancient Works, Monuments and Historical Sites: Preservation and Protection, 16-23, September 1988, Athens, P.G. Marinos and G. Koukis (eds), A.A. Balkema, Rotterdam, 1988, 707-716.
5. Price, C.A. Brethane stone preservative, BRE CP 1/81, Building Research Establishment, Garston, Watford, 1981, 2pp.
6. Building Research Establishment The selection of natural building stone, BRE Digest 269, Garston, Watford, January 1983.
7. Everett, A. Materials - Mitchell's building construction, B.T. Batsford, London, 1970.
8. Mitchell, G.A. and Everett, A. Building construction: Materials, B.T. Batsford, London, 1986.
9. Brown, E.T. (ed) Rock characterization, testing and monitoring: ISRM suggested methods, Pergamon Press, Oxford, 1981, 92-94.

10. American Society for Testing and Materials Standard method of test for soundness of aggregates by use of sodium sulphate or magnesium sulphate, ASTM C88-71, 1971, 49-53.

11. Sperling, C.H.B. and Cooke, R.U. Laboratory simulation of rock weathering by salt crystallization/hydration in a hot arid environment, Earth Surface Protection and Landforms, 1985, vol. 10, 541-555.

BIBLIOGRAPHY

Chadwick, S.J. The ageing of building stone, Unpublished Dissertation, MSc Advanced Course in Engineering Geology, University of Durham, England, 1991.

Dugdale, A.W. Durability of building stone, Final year BSc Honours Degree in Engineering Project Report, University of Durham, England, 1992.

Design principles of early medieval architecture as exemplified at Durham Cathedral

PROFESSOR E. C. FERNIE, BA (Rand), FSA, University of Edinburgh

SYNOPSIS. Despite the restrictions of the site the architect responsible for the original design and construction of Durham Cathedral made it one of the two or three largest churches of its generation, as befitted the ecclesiastical and secular status of its bishop. He was the first in Norman England to use spiral columns to mark sanctuaries, he established a workshop organization which produced the finest masonry in contemporary Europe, and he built the earliest high rib vaults north of the Pyrenees, though neither he nor any of his successors on the site invented the flying buttress.

INTRODUCTION
1. Where design is concerned we begin with a purpose, and principles then help direct our conduct towards achieving it. The goal to be examined in this case is the provision of a cathedral at Durham in the late eleventh century. As the House of God this structure had to provide a setting for the high altar and the shrine of St Cuthbert, room for the celebrants, the choir and the laity, spaces for subsidiary altars, security and protection from the weather for both people and furniture, and domestic buildings to house the monks who acted as clergy to the cathedral. It had also to compare favourably with the most imposing cathedrals and abbeys of Norman England, in order to maintain and enhance the prestige of the bishop in competition with his fellow ecclesiastics, and, perhaps more importantly, his standing as a temporal ruler, as a figure of secular power in a region which had been brutally pacified after the Conquest.
2. The principles governing contemporary ecclesiastical architecture on which there is widespread agreement among historians can be defined as follows.
(a) The design should in all essentials follow tradition while in true medieval fashion introducing

innovation wherever appropriate.

(b) Specifically, it should meet the requirements of church and state by following the tradition of the great church as developed in the Latin West between the fourth century and the eleventh.

(c) The design should be established using geometrical proportions selected to ensure both the stability of the structure and the harmonious interrelationship of its parts.

(d) It should use architectural means such as different types of columns, piers and shafts to express the significance of different parts of the building.

(e) It should employ carved decoration to enliven the surfaces of the stone and direct the eye in keeping with the overall design.

(f) The church should be built of stone and the quality of the masonry maximised through the use of the techniques of mass production.

3. There is one major difference of opinion on a point of principle, namely the extent to which plans were drawn up in advance of construction. One view holds that buildings were designed more or less in their entirety by a single architect, which is supported by, for example, the appointing of William of Sens to rebuild Canterbury Cathedral in 1174, or the succession of single architects named as being in charge of the rebuilding of Reims Cathedral after 1210. The other view contends that churches were built piecemeal as circumstances dictated, by a succession of masons designing their part of the building on the basis of what they found already built.

4. In fact both these views are correct, since procedures varied with circumstances. Buildings of great prestige where funds were available would tend to be designed in advance by one architect, whereas those of local status built on a very limited budget were more likely to be treated in the piecemeal fashion. Of course even the most fully prepared designs were often altered during construction following a change in taste or other circumstance, as indeed happened at both Canterbury and Reims.

THE DESIGNING OF DURHAM CATHEDRAL

5. All the evidence suggests that (with the exception of the vaults in the transepts and nave) Durham Cathedral was designed in or shortly before 1093 to look as it did on completion in 1133 (refs 13, 15).

6. The overall layout is one which had been standard since the time of Constantine's Christian Roman basilicas in the fourth century, that is an

aisled hall with an apse at one end. In the Carolingian period of the eighth to tenth centuries a transept and crossing were added, and this design passed with variations to the eleventh-century architecture of Northern France including Normandy. The most important churches built in England after the Conquest are larger than most of their counterparts on the Continent, and in almost every case during the 1070s each one is longer than its immediate predecessor, as column (i) in Table 1 illustrates. Winchester Cathedral, the largest Norman building of all, sets a standard for churches of the 1080s and 1090s. One would therefore expect Durham to be of a similar length to contemporary structures such as Bury St Edmunds Abbey, after 1081 (148m), or Canterbury Cathedral as extended after 1096 (133m), but it is restricted by its site to c. 117m, with only 15m between the façade and the cliffs to the west, and a similar distance between the apse and the falling away of the ground to the east.

Table 1. Lengths in metres of parts of some major Anglo-Norman churches. Column (i): total interior length to the head of the apse; (ii): distances between the eastern end of the nave and the chord of the apse.

		(i)	(ii)
Canterbury Cathedral	1070	84	23.25
Canterbury St Augustine's	1071	102	23.94
St Albans Abbey	1077	114	45.00
Winchester Cathedral	1079	157	33.59
Bury St Edmund's Abbey	1081	148	39.87
Durham Cathedral	1093	117	51.00
Canterbury Cathedral	1096	133	56.40
Norwich Cathedral	1096	132	31.62

7. In addition to vying with one another in terms of absolute size the great majority of large Anglo-Norman churches shared a system of proportions based on the relationship between the side of a square and its diagonal, that is 1:1.4142. Among many other aspects of the design this relates the side of the çcloister to the length of the nave. Durham follows this pattern: the north side of the cloister is 44.35m long which produces a diagonal of 62.71m, while the nave is 62.89m long to the centre of the façade wall (Fig. 1)(refs 11, 1, 8).

8. The likelihood that nave and cloister were intended to be proportionately related at Durham is supported by two observations. First, the bays of the nave vary considerably in length, increasing in size from east to west, with bays one to three each

DESIGN PRINCIPLES

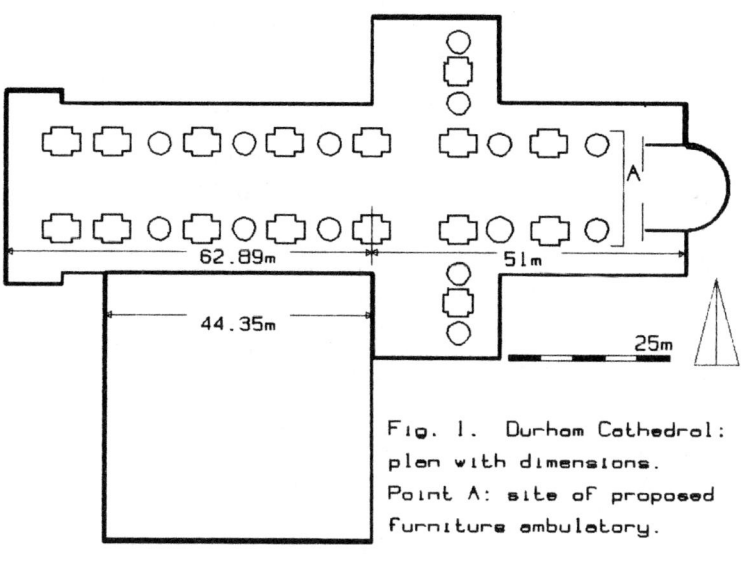

Fig. 1. Durham Cathedral: plan with dimensions. Point A: site of proposed furniture ambulatory.

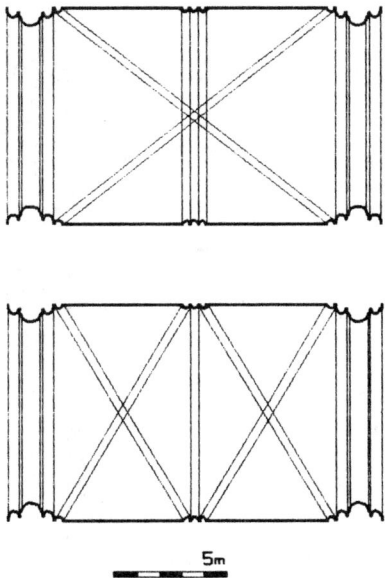

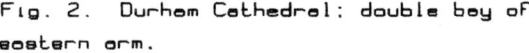

Fig. 2. Durham Cathedral: double bay of eastern arm.
 Top: with one six-part vault.
 Bottom: with two four-part vaults.

of 7.11m, four to six .3-.6m more than that and bay seven .91m more (bay 8 is almost 1.8m longer, but it contains the western towers). This proves that the nave was not laid out in a given number of identical units, and suggests the bays were on the contrary made to fit into a pre-determined overall length. The same conclusion is suggested by the fact that, in common with a number of buildings using this proportion, the west wall of the cloister does not meet the aisle wall in line with one of the piers as one might expect, but at an apparently arbitrary point in the seventh bay determined by the demands of the proportional system (ref. 7).

9. In the standard formula the length of the nave also relates to the distance between the west end of the nave and the chord of the apse in the same way, as 1:1.4142. A square with a side equal to the length of the nave of Durham would produce a diagonal of 88.94m, leaving 26.05m for the distance between the nave and the chord (that is the eastern arm without the apse, plus the east-west length of the crossing, which I shall refer to as the eastern section to distinguish it from the eastern arm). Durham however does not follow the rule as its eastern section measures about 51m, nearly twice as long as one would expect. The reasoning behind this may have run as follows. An overall length of 117m makes Durham a short building for its class and date, but in terms of its eastern section it emerges, after the extension to Canterbury Cathedral, as the second largest building of the Anglo-Norman period. The architect would thus have turned the restrictions of the site and the standard system of proportions to his advantage, and by combining this extensive eastern section with the massiveness of the piers and walls he has produced the powerful effect for which Durham is famous.

10. The siting of St Cuthbert's shrine reveals another important aspect of the way in which the designer approached his task. The east ends of large churches in Norman England take the form either of an apse flanked by smaller apses at the ends of the aisles, or an ambulatory continuing the aisles around the apse. In terms of gaining access to relics set in the usual place behind the high altar there is no doubt that the ambulatory is the sensible form to use, but at Durham the remains of one of the most important saints in the country were placed in what appears to be a dead-end, rendering them much more difficult to approach than they would have been from an ambulatory. The explanation for this arrangement probably lies in the provision in the original design of an ambulatory in the form of

DESIGN PRINCIPLES

wooden screens rather than as part of the structure of the building proper. This would have stood behind the high altar in the fourth bay from the crossing, giving access to the shrine to the east (Fig. 1, point A)(ref. 12).

11. This illustrates the importance of reading medieval buildings in terms of the way in which the masonry structure relates to the furniture which once made it usable. It is equally true that items of furniture can be appropriated into the architecture proper, as seems to have been the case with the columnar piers of the eastern arm and transept at Durham. That is, the spiral columns of the screen and ciborium of Old St Peter's in Rome have been incorporated into the alternating system of masonry columns and peirs at Durham in order to mark the eastern arm with the shrine and the high altar and the three altar areas in each arm of the transept (ref. 6).

12. The aisle on the eastern side of the transept is an example of a form designed for one activity, namely the channelling of processions, being used for a completely different one, namely the housing of altars. This function is suggested by the evident pointlessness of a processional aisle which ends at the end of each arm with no corresponding doorways, and by the spiral columns which mark the bays as sanctuaries and not just as aisle spaces beyond an arcade.

13. The fabric of the cathedral provides a number of indications of the extent to which stone-cutting was organised on an almost industrialized basis, involving preparation and standardization closely comparable with modern ideas of mass production. The following are three examples of this.(9)

14. Each course of stone in the building maintains the same depth, even where a run is over 120m long, as in the first phase in the eastern arm and transept. The application of this principle is expensive as it involves the loss of a great deal of raw material. Stone comes out of the quarry from beds of varying depth: if it is supplied and used without being reduced to batches of uniform measures the result will be a jigsaw such as is common in Armenian buildings of the first millennium, which are examples of a heroically direct approach to the mason's craft. If the stone is cut in sufficient quantities to provide a course of the same depth round a full circuit then this sort of mason's nightmare can be avoided, but the prerequisite is a very high degree of organization. That is, the position of almost every block of stone in the structure needs to be established in advance of the

start of building, and in the case of the lowest five courses in the first phase of building alone at Durham that amounts to over 1500 separate items.

15. Secondly, the position and size of the blocks making up the columnar piers have been arranged so that the patterns can be cut on separate stones on the workbench which are then assembled in the correct order during building. A very high degree of accuracy in producing identical blocks with identical markings is essential if mismatches across mortar joints are to be avoided. The demands are doubly great where the columnar piers have responds on their aisle faces, as with the four in the eastern arm, as these require the provision of yet more types of standard blocks (in addition to 920 of the curved type with diagonal incisions, there are 104 which are part cylinder and part respond, and 652 for the responds proper). Finally it is worth noting that the meeting point of column and respond lies 1.52m from the opposite face of the column's 7.13m diameter, two lengths which bear the same proportional relationship as that between the side of the cloister and the length of the nave. The Durham master handles all of this almost to perfection both at the level of geometry needed for planning and in the execution, to the extent that his achievements have been described as among the foremost intellectual developments of his time (ref. 5).

16. Thirdly, the arch mouldings are much richer and more complicated than those of the 1080s, heralding a new era in the decoration of buildings in the Norman tradition.

VAULTS

17. The eastern arm is now covered by a vault of thirteenth-century date, but the evidence supports the view that it was designed to be vaulted from the start. This evidence consists of (a) a full set of shafts which form an integral part of the gallery wall and which can have had no other purpose than to support a vault, (b) vault lines in the clerestory walls, (c) the lack of any sign that either the shafts or the webbing might have been intrusions, and (d) the lack of a clerestory passage which is present throughout the remainder of the building and is almost a rule in churches with any pretensions, the omission of which implies a wish for stability in the face of the thrust generated by the vault.

18. The form which the vault was planned to take is more controversial. John Bilson reconstructed two oblong four-part vaults over each double bay and considered a six-part vault out of the question.

DESIGN PRINCIPLES

More recently John James has pointed out that Bilson's elevation drawing shows no vault lines rising above the minor piers, making the six-part form more likely (refs 2, 10). The surviving masonry over the minor piers is ambiguous and if we used only this evidence the question would have to remain open. However, once one attempts to reconstruct vaulting on the supports available it is clear that their form must have been two four-part vaults over each double bay, as Bilson claimed. The argument runs as follows.

19. As there are no wall ribs in the other Norman high vaults in the building each of the ten shafts in each double bay must have supported either a transverse arch or a diagonal rib. In the case of a six-part vault the four corner shafts adjacent to the major piers would have supported the two diagonal ribs (Fig. 2, top). The two sets of three shafts over the minor piers could then only support a transverse arch of three elements, which would be almost as broad as the main transverse arch and hence very out of scale with the two diagonal ribs only one-third as thick. The result at the centre of the vault would have been extremely ungainly and would have divided the vault into two clear halves contradicting the six-part arrangement itself . With two four-part vaults the transverse arch over the minor pier would be a single rib like the diagonal ribs, creating a much more uniform distribution of rib forms across the whole double bay (Fig. 2, bottom).

20. In the transept, the south arm in its entirety and the north up to the base of the clerestory were designed and built without the intention of providing a vault. This is clear from the lack of any vertical provision for vaulting in the west walls of both arms, and from the form of the clerestory in the east wall in the south arm, where the triple arches of the wall passage all rise to full height, precluding the curvature of a vault of whatever type. The clerestory of the east wall of the north arm, however, was designed to accommodate a vault by making the flanking arches lower than the central one in each bay. As there is no parallel in this period for a church being designed with a vault over one arm of the transept and a wood roof without a vault over the other, one must conclude that the transepts were begun with the intention of vaulting neither arm, and that the south arm had been completed and the north arm constructed up to the level of the clerestory before it was decided to include vaults. At this point there was still an opportunity to adapt the design

of the clerestory in the north arm, but in the south the vault cuts across and obscures the flanking arches built without a vault in mind.

21. The vaulting in the north arm must have been built before that in the south because of the absence of chevron mouldings in the northern vault and their presence in the southern. Chevron mouldings do not occur at all in the eastern arm of the cathedral or in the transept at any level below the vaults. They first appear in the nave, in the third bay of the main arcade, the second of the gallery and the first of the clerestory, indicating that their introduction was a change of plan during building and not a predetermined piece of variety in the design. Thereafter they are ubiquitous, so that their appearance in the vault of the south arm clearly places its contruction after that of the north.

22. In the case of the nave, Bilson (in one of the few instances where his judgement has not been borne out by subsequent investigation) argued that the vault was intended, while Bony has shown convincingly that it was only incorporated into the design after the gallery had been built (refs 3, 4).

23. Bilson gives three chief reasons for his view. First, there is a narrow course under the corbels, suggesting provision for them in the building of the wall. There are however many such narrow courses bearing no relation to springing points. More important are the facts that no supports have been provided in the first place, making corbels necessary, and that the corbels and even the ribs themselves have been smashed into the wall in a manner which clearly indicates a change of intention. Bilson also points to a lack of alignment of mortar joints on either side of some of the corbels, suggesting the walls were built with them in place, but this is a misleading appearance due to the very uneven laying of the courses at this level. Thirdly, the clerestory windows are centred not on the the gallery arches but on the vaults. This only proves that the clerestory was built after the vault had been decided upon, not that the vault was part of the original design.

24. The nave vaults introduce two important innovations. Firstly the webbing decreases in thickness to the west, indicating an increasing understanding of the forces involved and the extent to which the structure could be pared away (ref. 3). Secondly the transverse arches are pointed, the earliest surviving examples used with a rib vault in England, probably deriving from those in the third church at Cluny in Burgundy begun in 1088 and coming

via the churches of the Cistercian order, the earliest English house of which was founded in 1128. At both Cluny and Durham the pointed arch appears to have been used for its practical properties as a more stable support than the semi-circular arch, rather than for the aesthetic qualities which dominated its use later in the century.

25. The half arches in the gallery of the nave are usually described as flying buttresses under the roof. Stephen Gardner has cast doubt on this assumption by calling attention to the extreme thinness of the original single arches of which they were composed. James has defended their status as buttresses by comparing their original section (.40m thick by 1.80m wide) with that of the flyers at Chartres (1.17m thick by .71m wide), giving them a similar area of cross section (.72 of a square metre at Durham and .83 at Chartres) But this calculation ignores the fact that the section at Chartres is more than 16% greater than that at Durham, while everything else at Chartres is proportionately dramatically thinner than its counterpart at Durham. Gardner instead sees the half arches as supports for the complicated roof over the gallery, which had a gable over the window in each bay, and compares them to the half arches at Norwich Cathedral, where there was no vault over the main span of the Norman building begun in 1096 (refs 9, 10).

26. To conclude, if at the start of the building of Durham Cathedral in 1093 only the eastern arm was intended to have a vault, then it must have been seen as a ciborium or canopy over the shrine and high altar, another example, with the spiral columns, of the appropriation of furniture into architecture. The extension of vaulting over the rest of the building therefore implies that the whole church had come to be considered as a ciborium, that is more like the Jewish Temple and less like the Early Christian meeting hall, reflecting one of the tensions evident in Christian church architecture from the very earliest days.

REFERENCES
1. BILLINGS R.W. Architectural illustrations of the cathedral church at Durham. London, 1843.
2. BILSON J. The beginnings of Gothic architecture. Journal of the Royal Institute of British Architects, 1899, 3rd ser., 6, 259-69, 288-324, 345-9.
3. BILSON J. Durham Cathedral, the chronology of its vaults. Archaeological Journal, 1922, 79, 102-60.
4. BONY J. Le projet premier de Durham: voûtement partiel ou voûtement total? Urbanisme et architecture: études écrites et publiées en honneur de Pierre Lavedan. Paris, 1954, 41-9.
5. BONY J. The stonework planning of the first Durham master. Medieval architecture and its intellectual context: studies in honour of Peter Kidson. Ed. E. Fernie and P. Crossley, London, 1990, 19-34.
6. FERNIE E.C. The spiral piers of Durham Cathedral. In ref. 14, 49-58.
7. FERNIE, E.C. Reconstructing Edward's abbey at Westminster. Romanesque and Gothic: essays for George Zarnecki. Boydell and Brewer, Woodbridge, 1987, 63-67.
8. FERNIE E.C. An architectural history of Norwich Cathedral. Chapter 4. OUP, 1993.
9. GARDNER S. The nave galleries of Durham Cathedral. Art Bulletin, 1982, 64, 564-79.
10. JAMES J. The rib vaults of Durham Cathedral. Gesta, 1983, 22, 135-39.
11. KIDSON P. Systems of measurement and proportion in early medieval architecture. 2 vols, PhD, London, 1956.
12. KLUKAS A. The architectural implications of the 'Decreta Lanfranci'. Anglo-Norman Studies, 1983/4, 6, 136-71.
13. MARKUSON, K.W. Recent investigations in the east range of the cathedral monastery. In ref. 14, 37-48.
14. Medieval Art and Architecture at Durham Cathedral. Transactions of the annual conference of the British Archaeological Association, ed. N. Coldstream and P. Draper, 1980.
15. SNAPE, M.G. Documentary evidence for the building of Durham Cathedral and its monastic buildings. In ref. 14, 20-36.

The unseen timber roofs of the choir, north transept, 1–2 The College and the dorter of Durham Cathedral

Eur Ing J. W. BULL, BSc, PhD, FIWSc, CEng, MICE, MIStructE, MIHT,
Lecturer in Structural Engineering, University of Newcastle

SYNOPSIS. This paper considers the use of timber in the roofs of the Choir, the North Transept, 1-2 The College and the Dorter of Durham Cathedral. Information on the historical records of the timber is given. A description is also given on the background to the timber engineering used in the construction of the cathedral. Sections through some of the roof timbers are given and a discussion on the dendrochronology on the roof timbers is included.

INTRODUCTION
1. A major problem faced by engineers wishing to research the cathedrals built in the Middle Ages, is that contemporary written descriptions of building methods are scarce. Facts on medieval design methods are lacking. There is also little written about how load transfer problems were solved. However, subtle problems like this have never stopped historians, so as an engineer I will continue.
2. In Saxon England, carpentry was the main building technique, so much so that when the Normans invited themselves to England, they had to import stone masons to build their minsters and halls. These stone masons used England as a test bed for many of the structures later built in France. Durham Cathedral is one of the test bed structures and is the least altered of the Norman churches in England, remaining almost as it was in the 12th century [1,2].
3. This paper does not deal with the stone structure, but with the now much neglected use of timber for structural engineering purposes. It should be remembered that even as late as the end of the 19th century, more bridges were constructed using timber than from any other constructional material. After all there is little engineering difference between using timber to span the north transept of Durham Cathedral and using the same timber to build a bridge of similar span.

THE BUILDING SEQUENCE
4. Between 1093 and 1104, the high vault of the choir together with its aisles, but with the exception of the easternmost bay and the north and south transepts, were completed [2,3]. The

vaulting required considerable amounts of timber centring. On completion of the vaulting, the timber would then have been used for other purposes.

5. Between 1110 and 1138 the north transept was vaulted and the vaulting of the nave and its aisles were completed. Between 1175 and 1185 the Galilee chapel was covered with a timber roof, but by 1428 that roof was in a ruinous state [2].

6. The evidence given in a lawsuit about 1225 shows that the Prior's lodging, which was under construction between 1195 and 1208 had timber provided by the bishop [4]. The timber for the western towers, completed about 1225 was brought from beyond the River Derwent, in this case from the Redleyhope area [4].

7. In January 1249, Bishop Farnham granted permission to appropriate the great tithes of the parish of Bedlington partly for the maintenance of the roof when it was finished [4,5]. [The author of this paper is a churchwarden at Bedlington]. In 1254 the timber of a wood near Brancepeth was leased for thirty years and it is believed used for the roof of the Nine Alters transept [4,6].

8. In 1398 the dorter [The Great Dormitory] was transferred to the western side of the cloister. This move required considerable rebuilding as the western side of the cloister was badly roofed and the rain had rotted and weakened the timber [4,7]. For the roof, standing timber from Redleyhope and timber purchased from Barnard Castle was used [4]. Contracts for the dorter are dated 22nd September 1398 and 2nd February 1401.

9. The cloister was virtually complete by 1419 with low-pitched roofs and flat wooden ceilings. The timber came from Baxter Wood, Bearpark, Rainton and Shincliffe. Imported Norway timber planks were purchased at Hartlepool and Newcastle, with Kendal boards of oak coming from beyond the Pennines [3,4]. The cloister is today substantially as it was then, despite 19th and 20th century restorations. For example, the roofs of the Cloisters were restored in 1828 [3].

10. During a storm in 1429 the timber belfry of the central tower was set alight by lighting. By 1456 the timber in the tower was in a state of great decay, but in 1459, the tower was again set alight by lightning with the roofs being badly burnt [4,7]. Between about 1455 and 1495, the central tower was rebuilt above the roof level of the cathedral.

11. Much of the roof on the south side of the cathedral was repaired in the early 1430's [4]. In the period 1429 to 1435 restoration of the Galilee Chapel records the use of timber from Bearpark and planks being purchased from Hull and Newcastle [4].

12. The East Range of the College at Durham Cathedral is dominated by the College Gate and No 3 the college. The College Gate may have occupied this site from early medieval times, but the present College Gate was built in the early 16th century and contains its original 16th century timber structure. Nos 1 and 2 The College, are between the College Gate and No 3 and formed part of a large hall or barn-like structure. The original purpose of Nos 1 and 2 is not known, but possibly were stables [9].

TIMBER ROOFS

13. Until about 1657, the two western towers were surmounted by tall timber spires covered with lead [3].

14. Between 1798 and 1803 there was considerable work on the main roofs. The entire ancient oak high pitched nave roof was replaced with the present lower pitched roof and the ridge line brought down to match that of the choir. The original gable apex now standing some three meters above the new ridge. For the choir roof, the old trusses and their purlins are in their original positions, but the heavy oak common rafters were removed and replaced by light softwood spars [10].

15. Earlier in this section, reference is made to timber being purchased from the towns [ports] of Hartlepool, Hull and Newcastle. This does not necessarily mean that the timber came from overseas as much of the internal English trade was transported by ship. The road system being less than adequate at the time!

ENGINEERING

16. Throughout the Romanesque period the pitch of nave roofs increased. By the middle of the 12th century, the roof pitch was 45 degrees, which required the tie beams to be of large section, with vertical members going up to support the roof. The tie beams were also now being used as platforms and lifting systems for the vaulting.

17. On completion of the walls, the timber roof was built and the vaults then constructed. The timber roof provided weather protection to the scaffolding and to the centring, plus preventing the rapid deterioration of the extensive falsework used in constructing the high vaults. The construction of the vaulting would have taken many years and the work would not have been completely continuous [11]. The rate of progress being governed, in the main by financial rather than technical considerations, but technical precautions were taken to allow the vaults to bed in before the centring was taken down. The cathedral would have been in use despite the continuing construction of the vaulting

18. It can be assumed that some sort of design and a plan of the cathedral would have been drawn up for construction purposes. The cathedral builders did not have photocopiers, or even persons skilled in the art of quickly and accurately copying and updating the engineering drawings. There would have been an original plan and probably no copies. Any plan that did exist would have been vulnerable to loss. Let us not forget that Gaudi's "Temple Expiatori de la Sagrada Familia" started in 1882, had all its plans and nearly all of its models destroyed during the Spanish Civil War, but its construction continues.

Analysis

19. The early part of the 12th century was an age of scientific and architectural relearning, when Western European scholars were beginning to obtain translated Arab texts on mathematics and science. Many texts were translated from the Greek into Syriac, then into Arabic and finally into Latin. However, many of the

original Greek texts were lost.

20. The cathedral builders were illiterate, innumerate and unlikely to have themselves left written records. They were not concerned with theorems, proofs or scientific validity; they were concerned only with the stability of the structures they built. They would commence construction with a bay span or a width and generate the remaining dimensions by applying numerical rules and constructive geometry, never resorting to calculation. We can also assume that the medieval builder had no means of calculating the stresses in the roof timbers. The section sizes used in Durham Cathedral are far larger than those required today.

21. For medieval building purposes, numerical rules were related to the square and to the compasses. Set ratio's for some triangles were used as they were simple and could be interrelated without the use of calculation. Trial and error construction was an alternative. Perhaps much of medieval building came down to the construction of a model and then model analysis, which when related to relative proportion, was scaled up and linked to the idea of stability. If it stands up it is too strong, if it falls down it is too weak. In the Middle Ages, the collapse of a structure was accepted as part of the "construction process learning curve," but this is not the case today!

Construction

22. During the medieval construction process, to determine the end positions of the roof timbers, that is, to measure distances, some form of rod or cord would have been used. The actual distance in units of measurement being meaningless to the majority of the builders. Usually a full scale drawing of the required roof timber would be made and used as a template. The roof sections would be cut, jointed and test assembled on the ground. The sections were then disassembled, hauled up to roof level and reassembled [11]. The use of this full scale three dimensional model building kit meant that the medieval builders worked at a slow pace using practical experience and on the spot supervision.

23. The use of actual dimensions rather than plans, was prudent as many cathedrals had inadequate foundations which allowed walls to be other than vertical. For example, the one degree rotation of a twenty metre high wall, would require a change in the length of a roof tie beam of 350mm. Further, the builders used setting out tolerances that would not be accepted today. Well, after all, if the axis of a cathedral had a bend in it, the builder could put in a screen at the bend point to break up the line of sight! If a column is incorrectly located, then the length of the roof timbers is also effected. Would the column locations in the nave be acceptable to todays builders? I doubt it. Also, the relationship between consultant and contractor did not then exist as they were essentially the same group of persons [Mistakes, what mistakes! We designed it that way!]. Fortunately, much of the foundations of Durham Cathedral are carried down to the bedrock and it is only the setting out tolerances that might need to be considered.

TIMBER ROOFS

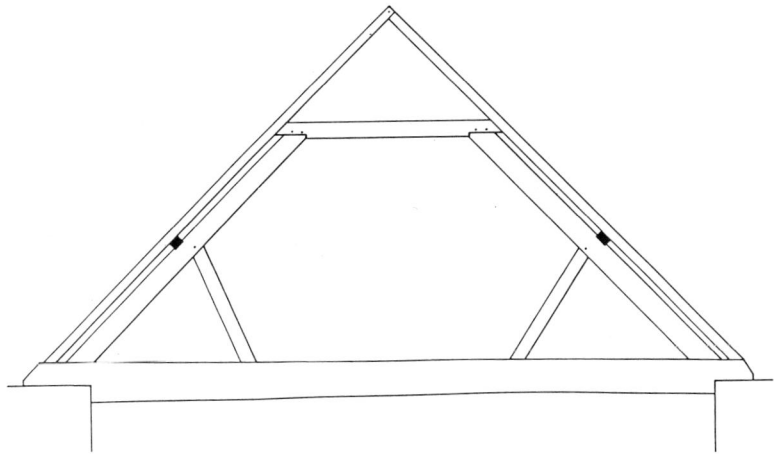

RCHME Crown Copyright
Fig. 1 Section through the north transept roof. Scale 9mm = 1m.

Practical considerations

24. Roof timbers have to resist the self weight dead load due to the whole roof structure, plus the live loads due to wind, snow etc. Timber has an advantage over stone in that it is strong in compression, strong in tension and can sustain high stress reversal loads due to wind. Stone structures can resist compressive loading but not tensile loading. The main engineering problem for the medieval builders was to produce a vault always loaded in compression, a timber roof resisting both compression and tension with the vault separated from the roof.

25. The timber roof would form a frame and have a weatherproof covering of lead or shingles. The medieval builders did not use the triangle to insure structural rigidity; they used a rectangular framework stiffened with the aid of angle brackets, braces and struts [11]. To increase member stiffness they would increase the depth of the principal rafters [11]. The feet of the rafters would be jointed into wall plates, which were mounted on the walls on either side of the span and tied across to resist the tendency of the walls to move horizontally.

26. In essence the medieval timber roof, if correctly designed was a self-restraining structure dependent upon the efficiency of the joints [12]. From a practical point of view, the length of a piece of timber is restricted to about 12 meters. For longer lengths two or more pieces were joined. The joints were kept as far as possible in compression. This was due to the essential weakness in tension of the pegged pin joint, the mortise and tenon joint, the tongued joint, the slot and tenon joint and the halved or the notched joint [11]. The fire resistance of a compression joint was also higher than for a tension joint; fire being an ever present problem.

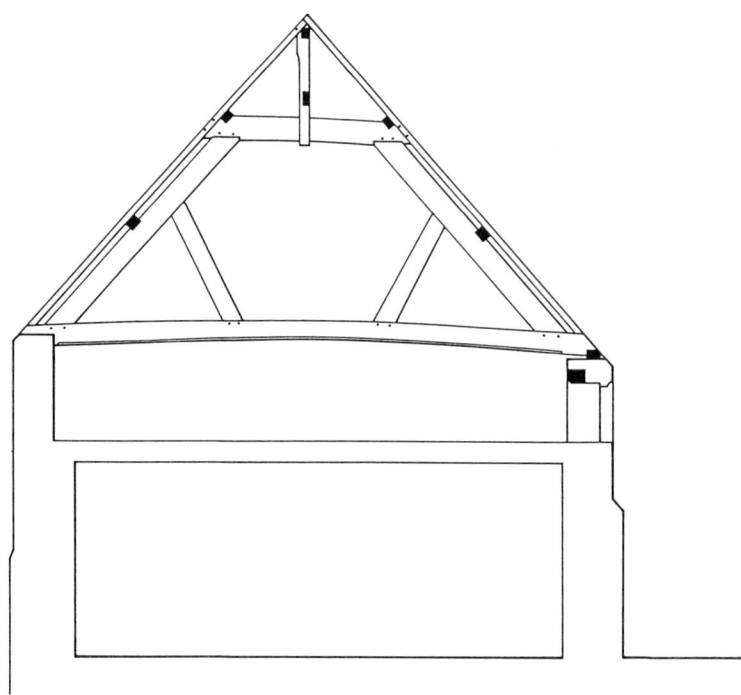

RCHME Crown Copyright
Fig. 2 Section through 1-2 The College. Scale 1cm = 1m.

DENDROCHRONOLOGY
27. Dendrochronology is the fixing of past dates by a comparative study of the annual growth rings in ancient trees. Put very simply, each year a tree produces a new layer of wood just under the bark. The growth of the wood starts in the early part of spring and stops in the winter months. This growth process results in clearly visible concentric rings known as annual rings. Each ring represents a year and hence the age of the tree is known. In Europe, as in many other non-tropical areas of the world, the annular rings show up clearly as two layers, an inner layer of large cavities called springwood and an outer layer of small cavities called summerwood. Variations in climate also affect the way in which the two layers grow. In tropical regions of the world, trees grow throughout the year and the two layers are seen as almost one layer. If a tree has a known age, then the character annular rings for specified years in a particular part of the world can be obtained. The annular bands nearest to the bark are called sapwood and the central core of the tree is called the heartwood. Because sapwood is not as resistant to fungi attack, roof beams use as much heartwood as possible. If bark is present on the beam, then the date of felling of the timber can be determined.
28. In Ireland, it is possible to go back about three thousand

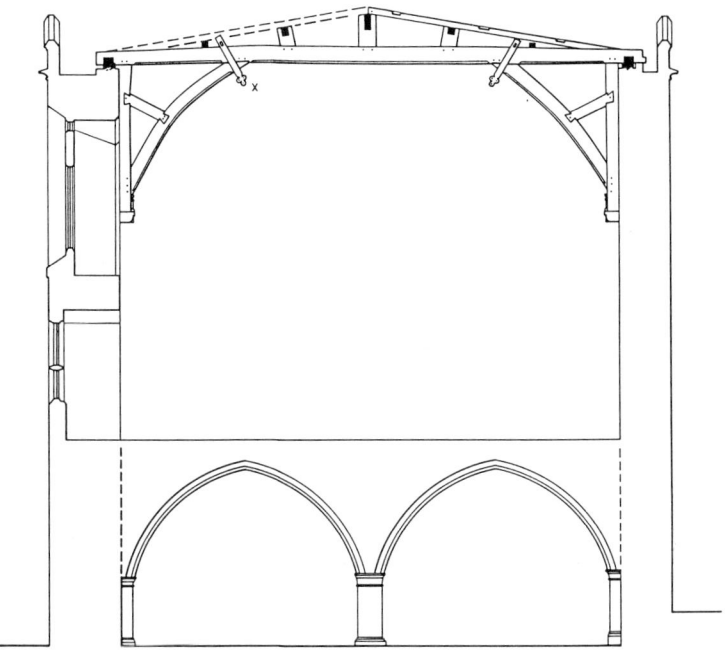

RCHME Crown Copyright
Fig. 3 Section through The Great Dormitory at the fourth truss, looking north. Scale 6.5mm = 1m.

five hundred years with the dendrochronology data. However, the timber samples from Durham Cathedral were related to the data for the East Midlands, for which accurate records from 822 to 1981 are available. Consequently, if a series of annular rings are available on say a roof beam, then by comparing the rings with confirmed records, the date of the growth of the timber and the area of the world it was grown can be determined and a guess made at the date of construction of the roof. For example, in one sample of timber, many hundreds of years old, it was possible to determine the month in which the tree was felled. A second example concerned a timber ship, again many hundreds of years old. The ship was found to be constructed of timbers grown hundreds of miles from where the ship came to rest. Perhaps the ship's navigation was unsatisfactory or European trading was far more extensive than historians realised!

29. The dating records given in the following tables come from the Royal Commission on the Historical Monuments of England and were determined by the Nottingham University Tree-Ring Dating Laboratory [13]. They were sampled on the 11th and 12th October

1990. In total twenty four samples were taken and twenty one analysed. Tables 1, 2 and 3 show the dates of growth of the trees using the following notation:-
b = bark was present on the timber but not on the sample,
C = complete to bark on the sample,
c = complete to bark on the timber but not necessarily on the sample,
H/S = heartwood/sapwood boundary,
NM = not measured and
.5 = part of a growth ring was present but not measured.

TABLE 1. Durham Cathedral Choir Roof

Sample Location	Total Rings	Sapwood Rings	First Dated Ring	Last Heartwood Ring	Last Dated Ring
Truss 12 Tie Beam	86.5	15.5C	1372	1442	1457
Truss 13 Tie Beam	113	18	1346	1439	1458
Truss 14 Brace South Side	102.5	32.5C	1356	1425	1457
Truss 15 Tie Beam	82	H/S	1352	1433	1433
Truss 16 Truncated Principal South Side	86	8.5	1370	1447	1455

30. From Table 1 trusses 16, 15, 14 and 13 have a sequence of 113 rings and are dated to 1346 to 1458. Further, trusses 12 and 14 appear to have a complete sapwood to the bark. If this is the case, then trusses 12, 14, 15 and 16 were felled in the period 1455 to 1458 or very soon afterwards. The date of felling of truss 13 is not known.

TABLE 2. Durham Cathedral North Transept

Sample Location	Total Rings	Sapwood Rings	First Dated Ring	Last Heartwood Ring	Last Dated Ring
1. Tie Beam Truss 2	145+42NM	H/S+42NM	1534	1678	1678
2. Tie Beam Truss 3	191	25C	1538	1703	1728
3. Tie Beam Truss 4	156+20NM	6+20?b	1561	1711	1716
4. Brace East Side Truss 4	70	23C	1388	1434	1457
5. Tie Beam 6	151	32C	1564	1682	1714
6. Brace East Side 6	87	14	1363	1435	1449
7. Tie Beam 5	137.5	26.5	1320	1430	1456
8. Truncated Principal East Side Truss 3	83.5	22.5	1372	1432	1454

31. From Table 2 samples 1, 2, 3 and 5 have a sequence of 195 rings and are dated to 1534 to 1728. Also samples 4, 6, 7 and 8 have a sequence of 138 rings and are dated to 1320 to 1457. Further, sample 2 has complete sapwood giving a felling date of 1728/9 and sample 5 also has complete sapwood that gives a felling date of 1714/5. Samples 4 and 8 have complete sapwood suggesting samples 4, 6, 7 and 8 were felled in the period 1454 to 1458.

TABLE 3. 1-2 The College, Cathedral Precinct, Durham Cathedral

Sample Location	Total Rings	Sapwood Rings	First Dated Ring	Last Heartwood Ring	Last Dated Ring
1. Tie Beam Truss E	163	19C?	1369	1512	1531
2. Tie Beam Truss D	125.5	29.5C?	1406	1501	1530
3. Tie Beam Truss C	152	28C?	1380	1503	1531
4. Tie Beam Truss B	156	22C	1376	1509	1531
5. Truncated Principal West Side Truss C	153	4	1364	1512	1516
6. Truncated Principal East Side Truss C	151.5	21.5C	1381	1510	1531
7. Truncated Principal East Side Truss B	149	H/S?	1364	1512	1512
8. Truncated Principal West Side Truss B	102	H/S?	1405	1506	1506

32. All the samples in Table 3 form a site sequence that has 168 rings and are dated to 1364 to 1531. Samples 1, 2, 3, 4 and 6 are thought to have complete sapwood and would suggest that the trees were felled in 1531/1532.

33. Each table has a sequence of rings. This indicates, that the planting of trees was a continual process, such that as trees were felled, they were replaced.

ROYAL COMMISSION [FIGURES 1, 2 AND 3]

34. A series of surveys of the roofs of Durham Cathedral has been undertaken by the Royal Commission on the Historical Monuments of England [13]. Not all the surveys have been completed, but those that have been completed and relate to the timber roofs are included. Fig 1 shows a section through the North Transept roof and is complementary to Table 2. Fig 2 shows a section through the roof of 1-2 The College and is complementary to Table 3. Fig 3 shows a section through The Great Dormitory, at the fourth truss from the south end, looking north. As yet the Great Dormitory has not been dated by dendrochronology. The drawings for the Choir roof are yet to be completed.

BULL

CONCLUSIONS

35. As the timber samples were related to, and found to be compatible with, the East Midlands data, this suggests the timber was English and not imported from other countries. It could also be inferred that much of the English internal commerce was transported by ship rather than by road.

36. Throughout the first half of the 15th century, there was a constant stream of repairs and renovations to the cathedral and items of expenditure continue to appear in the accounts for the rest of the Middle Ages. The only major item of expenditure during that time was on the central tower, which by 1456 required restoration. The Prior sought the advice of the Bishop in 1456. On Easter day in 1459, the tower was struck by lighting and set alight [3,4]. Records speak of damage to the nave, the tower, aisles and other buildings [4,8]. If the tower were set alight, it is reasonable to assume that large areas of the roof of the nave, the choir and the north and south transepts and aisles would have been damaged. Records show that work was in progress on timber construction in 1465/6 and that repair work continued to the late 1480's.

37. Table 1 indicates that the timber for trusses 12, 13, 14, 15 and 16 in the Choir roof was felled between 1455 to 1458 or very soon afterwards. Perhaps the letter to the Bishop in 1456 prompted the felling of the trees and plans were in hand to rebuild the tower. If the tower was in such a state of disrepair as the records suggest, then timber for the roofs would also be required. The fire of 1459 extending the area requiring repair.

38. Table 2, which refers to the North Transept roof includes timber felled in the period 1454 to 1458 and used for trusses 3, 4, 5 and 6. This again indicates the extent of the repairs due to the tower and to the fire of 1459.

39. The timber described in Table 3 for 1-2 The College, is dated to 1364 to 1531 and samples 1, 2, 3, 4 and 6 suggest that the trees were felled in 1531/1532. The present College Gate, adjacent to 1-2 The College was built in the early 16th century. Table 3 suggests that the roofs of 1-2 and perhaps the College Gate were constructed during the mid to late 1530's.

40. In the Middle Ages, when timber was felled, it was usually pit sawn to the approximate shape and then left to dry out or season. Seasoning preventing splitting of the timber when in structural use. For the timber described in Tables 1, 2 and 3, seasoning would have been accomplished by possibly immersing in water, then always by air drying. The process taking a number of years and ensuring that unless records were kept, the precise date of the roof repair would be unknown. Hence the interest in samples 1, 2, 3 and 5 of Table 2. Sample 5 has a felling date of 1714/5 and sample 2 a felling date of 1728/9. This would suggest that repairs to the North Transept roof were taking place in the 1710's to the 1730's. There do not appear to be any records of these repairs [13].

41. A further point regarding the tables, is that within each

table the felling dates are close together. This suggests that or the Choir, the North Transept and 1-2 The College, the timber was felled to order. It is more usual for felling to proceed at a steady rate and the felled timber to be stored. The stored timber acting as a buffer between supply and demand. However, if the restoration of the central tower and the roofs during and after the period 1456 to 1459 was more extensive that thought, then having used the stored timber, felling to order may well have meant using inferior quality timber. This may go some way to explaining the repairs to the North Transept in the 1710's to the 1730's and to the change in roof pitch of the Nave and the Choir at the end of the century.

42. When carrying out a dendrochronology survey of the cathedral roofs, it is essential that the precise locations of the samples are recorded. This requires the accurate surveying and the drawing of relevant plans. Although the Royal Commission have not yet completed these drawings, they have found significant differences between the actual roof structure they surveyed and "plans" published in the last century or so. Thus it was not only the medieval builders who had problems getting their plans right!

43. When the Royal Commission have completed their drawings and a full dendrochronology survey of the cathedral roofs have taken place, it will be possible for a history of the timber to be drawn up. This history could then be used to fill in the gaps in the written historical records of the cathedral and where records are available, to determine the place of felling of the timber plus the lead time and planning required to construct the timber roofs. Perhaps the full extent of the repairs and of the fire of 1459 can then be determined, although this may not be possible on the roof of the Nave as its pitch was reduced between 1789 and 1803.

ACKNOWLEDGEMENTS

44. I would like to acknowledge the immense help given to me by the Royal Commission on the Historical Monuments of England, especially; a) Robert Hook, Head of Threatened Buildings [North], Shelley House, Acomb Road, York, YO2 4HB and b) Ms. Sarah Brown, on the staff of the National Building Records, Fortunes House, 23 Saville Row, London, W1. [Tables and Figs. 1, 2 and 3 are from the Royal Commission]. I also acknowledge the help I received from Cannon R. Coppin, Mr. I. Curry and Mr. C.J. Mettem. I would also like to thank the Rev. Dr. M.J. Jackson for all his efforts.

REFERENCES

1. Harvey J. The Gothic World 1100-1600, R. T. Batsford Ltd, London, 1950.
2. Cook G.H. Portrait of Durham Cathedral, Phoenix House, London, 1948.
3. Boyle J.R. The County of Durham, its castles, churches and manor houses, W. Scott, London, 1892.

4. Snape M.G. Documentary Evidence for the Building of Durham Cathedral and its Monastic Building, The British Archaeological Association, Vol III, Medieval Art and Architecture at Durham Cathedral, Conference Transactions for 1977, 1980, 20-36.
5. Luard H.R. Matthei Parisiensis Chronica Majora, Rolls Series, LVII,V 1880, 10; Miniments of the Dean and Chapter of Durham, 2.1.Point.14.
6. Miniments of the Dean and Chapter of Durham, 2.11.Spec18, printed Script. Tres, Appendix lxxxiii, LXVII. Cf. Greenwell, 79n.
7. Account Rolls, 585, Chapters of the English Black Monks, III, Camden Society, 3rd Ser., LIV, 1937, 84.
8. Snape M.G. Durham Cathedral: an Unknown Fire, TAASDN, New series, III, 1974, 71-74.
9. Curry I. Report on the east range of The College, Durham Cathedral, 30th December 1989.
10. Curry I. Restorations and Repairs to the Fabric of Durham Cathedral 1777 to 1803, The British Archaeological Association, Vol III, Medieval Art and Architecture at Durham Cathedral, Conference Transactions for 1977, 1980, 130-139.
11. Fitchen J. The Construction of Gothic Cathedrals, Oxford University Press, 1961.
12. Morris R. Cathedrals and Abbeys of England and Wales, The building church, 600-1540, J.M. Dent & Sons Ltd, London, 1979.
13. Hook R. Royal Commission on the Historical Monuments of England, private communication, December 1992.

The roof of the monks' dormitory, Durham

J. HEYMAN, FSA, FICE, FEng, Emeritus Professor of Engineering,
University of Cambridge

SYNOPSIS. The main beams of the early fifteenth-century roof of the Great Dormitory at Durham span over 12 m. They are supported on the external walls and also on knee braces, so positioned that the design is economical and stresses are low.

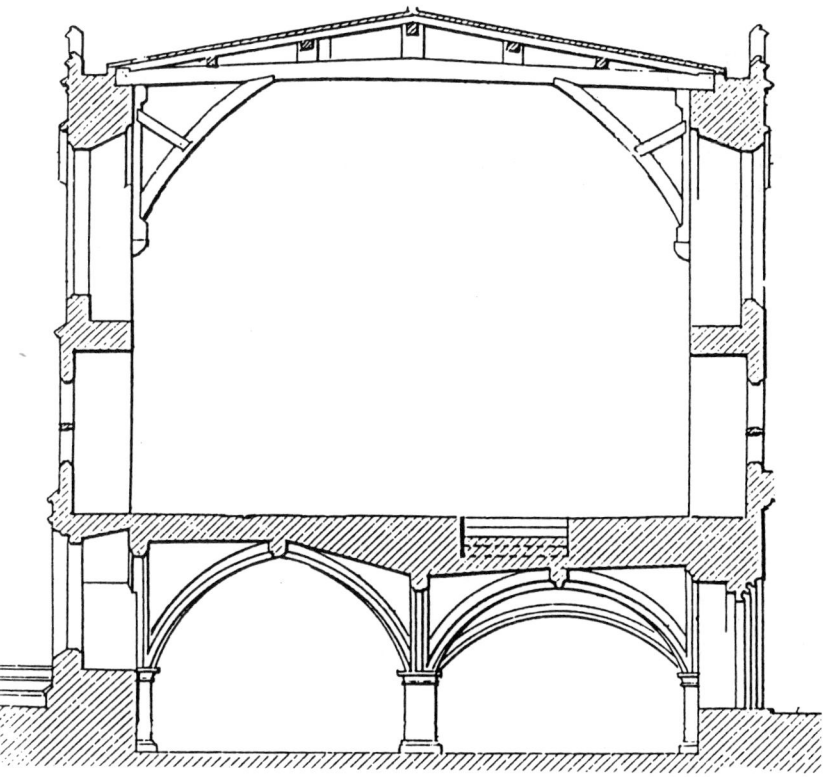

Fig. 1   The Monks' Dormitory

Engineering a cathedral, Durham. Thomas Telford, London, 1993

INTRODUCTION

1. Figure 1 shows a cross-section of the Great Dormitory of Durham Cathedral. The stone undercroft was vaulted in the mid-thirteenth century; the upper chamber, 40 ft wide, 30 ft high and 194 ft long (12.3 m × 9.2 m × 60 m) was roofed early in the fifteenth century. The twenty-one great oak trees forming the roof beams are spaced at 2.8 m centres, Fig. 2.

2. The roof system is simple. The lead cladding is carried directly on boards about 25 mm thick, and these transfer their loads to the common

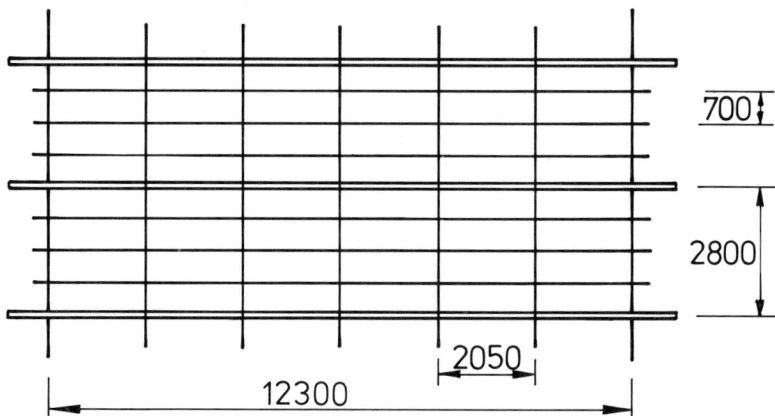

Fig. 2   Schematic plan of roof

rafters spaced at 700 mm. The common rafters are supported in turn from the side walls and from the ridge purlins, with intermediate purlins transmitting their loads through short posts to the great beams.

3. Thus, in an idealised way, each great beam "looks after" an area of roof 12.3 m × 2.8 m, and is subjected to a series of point loads whose

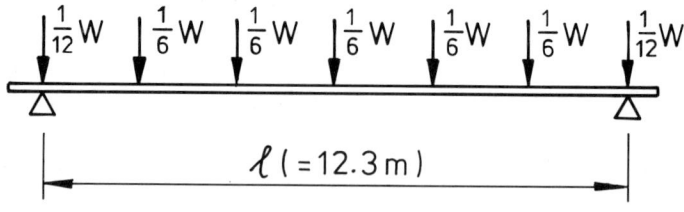

Fig. 3   Loads on roof beam from purlins

total is W, Fig. 3. The value of W corresponds to the load being supported by the beam, and it is of interest, before proceeding further, to give some idea of its magnitude. Loads used in the calculation were:

Weight of timber	0.95 kN/m²
Weight of cladding	0.65
Total dead	1.60
Superimposed (snow)	0.75
Dead and superimposed	2.35 kN/m²

The value of W (dead and superimposed) is therefore $(2.35)(12.3)(2.8)$ = 81 kN. If the beam in Fig. 3 were truly simply supported, then the maximum bending moment would have value

$$\frac{W\ell}{8} = \frac{(81)(12.3)}{8} = 124.5 \text{ kNm}.$$

4. The beams have cross section about 320 mm × 485 mm, with a corresponding elastic section modulus of 12.55×10^6 mm³, so that the corresponding stress in bending is 9.9 N/mm². This is perhaps high as a secular value, although it is certainly less than the range of grade stress in bending permitted by British Standard 5268: Part 2: 1984 for hardwoods.

5. Conventional assumptions are of course hidden in even these simple calculations. For example, it is usual to assume that apparently similar members will behave similarly; in particular, a uniformly applied load will be shared equally by the members concerned, leading eventually to the fact that the point loads noted in Fig. 3 have equal values. Examination of the roof at close quarters shows immediately that there are anomalies. As an instance, some of the common rafters are not actually in contact with their "supporting" purlins at all points within their lengths, so that all loads are certainly not shared equally. However, the "equal-sharing" assumption of conventional calculations is in reality safe. This statement may be justified by detailed argument; in descriptive terms, an overloaded portion of a structure will give way and force a correspondingly underloaded portion to share in its proper task.

6. Thus the very simple "simply-supported-beam" calculations indicate that the 21 great beams are of sufficient size to support the required loading. However, the roof system sketched in Fig. 1 is more complex, in that the ends of each beam are supported not only on the side walls but also on knee braces installed at roughly 45°. The presence of these braces will reduce markedly the stresses in the main beams, and will also reduce deflexions.

THE BEAM ON FOUR SUPPORTS

7. In Fig. 4 the raking braces are represented by two extra support forces S, in addition to the forces R provided by the side walls.

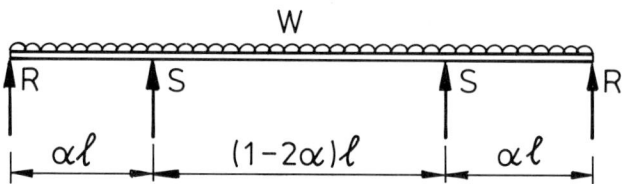

Fig. 4   Beam on four supports

By comparison with Fig. 3 it will be seen that the (already idealised) point loads have been replaced by a uniformly distributed load of the same total value; again, this is in accordance with usual procedures of structural analysis. (An excursion is made below in order to indicate the numerical differences that are introduced by this distribution of the point loads. The errors are small, and it may be noted that the "free" bending moment $W\ell/8$ at the centre of span is identical for the systems of Figs 3 and 4.)

8. The supports from the braces are indicated in Fig. 4 as placed at a distance $\alpha\ell$ from the walls, and this paper is concerned with the consequences of varying the value of α. It may be imagined that for small values of α (shorter braces than those shown in Fig. 1) the ends of the great beams will lift off the external walls, so that each beam is supported only by the two forces S in Fig. 4, no reaction being supplied by the walls. As will be seen, the braces are actually positioned so that this situation could arise, and there are indeed indications from a close examination of the dormitory roof that the walls are in some instances providing little support.

(a) No support from the walls; $\alpha < 0.2142$

9. In Fig. 5 the beam is supposed to have lifted from the walls; the corresponding bending-moment diagram has two cardinal values, M_s and M_c. The values of these bending moments are

$$\left. \begin{array}{l} M_s = (2\alpha)^2 \, \frac{W\ell}{8} \\ M_c = (1 - 4\alpha) \, \frac{W\ell}{8}. \end{array} \right\} \quad (1)$$

10. It is a standard undergraduate problem in the theory of determinate structures to find the "optimum" position of the supports S, that

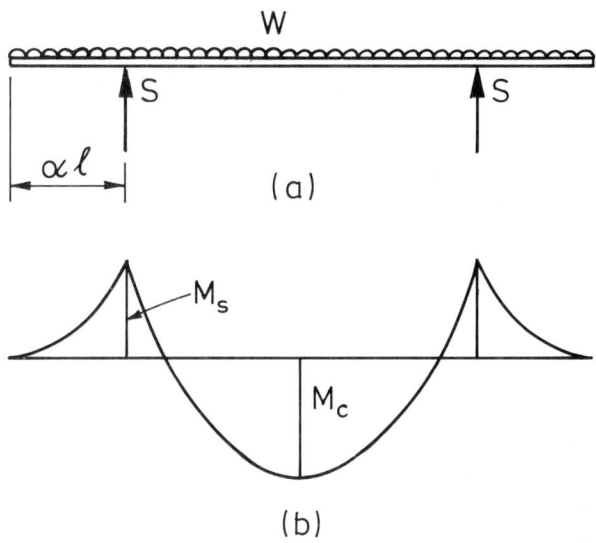

Fig. 5   Beam on two supports

is, to find the value of α for which the values of M_s and M_c are equal. Equating the two expressions in equations (1) leads to

$$4\alpha^2 + 4\alpha - 1 = 0$$
$$\text{or } \alpha = \tfrac{1}{2}\left(\sqrt{2} - 1\right) = 0.2071. \tag{2}$$

For this value of α,

$$M_s = M_c = (0.1716)\frac{W\ell}{8}. \tag{3}$$

The condition of lift-off holds, as will be seen, for $\alpha < 0.2142$; above this value, the reactions R of Fig. 4 help to support the beam.

11. The analysis may be repeated with the slightly less-idealised assumption of equal point loads, Fig. 6. Only the case $\tfrac{1}{6} < \alpha < \tfrac{1}{3}$ (as sketched in Fig. 6) will be considered. The cardinal values may be calculated as

$$\left. \begin{array}{l} M_s = \left(2\alpha - \tfrac{2}{9}\right)\frac{W\ell}{8} \\ M_c = (1 - 4\alpha)\frac{W\ell}{8}. \end{array} \right\} \tag{4}$$

Equating these two expressions gives

$$\left.\begin{array}{l} \alpha = \frac{11}{54} = 0.2037 \\ \text{and } M_s = M_c = \frac{5}{27}\frac{W\ell}{8} = (0.1852)\frac{W\ell}{8}, \end{array}\right\} \quad (5)$$

cf. the values of the equations (2) and (3).

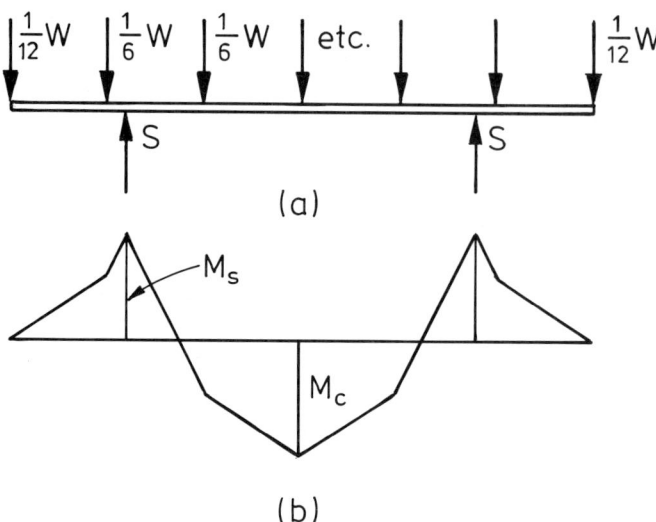

Fig. 6   Beam with point loads

(b) Support from the walls; $\alpha > 0.2142$

12. The general bending-moment diagram for the roof beam on four supports is sketched in Fig. 7. The beam has, of course, a single redundancy; that is, the values of R and S must together total $\frac{1}{2}W$, but there

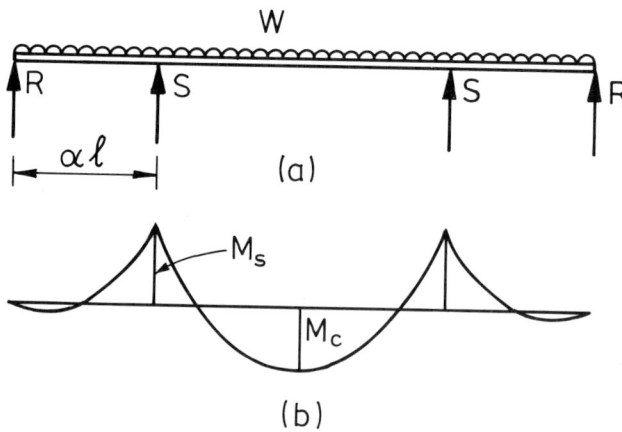

Fig. 7   Beam on four supports

is no other relationship between these two quantities that can be found from the simple laws of statics. To find their values, a full application of the theory of structures must be made.

13. The results of a conventional elastic analysis will be presented first, but it will be appreciated that several more or less doubtful assumptions are built into such an analysis. For example, unless the working is to become excessively tedious, the beam must be assumed to have uniform cross-section throughout its length; moreover, the basic elasticity of the wood will also be taken as uniform. Both of these assumptions are clearly only very approximate representations of the roof at Durham.

14. However, more critical and important assumptions arise in the specification of the boundary conditions for the elastic solution; some decision must be made about the possible displacements of the four support points R and S in Fig. 7. The results quoted below have been obtained by assuming that all supports are rigid (or, more precisely, that they stay all at the same level), whereas a very small differential settlement will have a marked effect on the values of the bending moments in the beam. However, no other assumption is really possible, since the support conditions are, in essence, unknowable, depending as they do on such factors as compressibility of the angled braces, shrinkage of connexions between the timbers, possible decay of supporting wall plates, and so on.

15. The following values which result from the elastic analysis must therefore be viewed with some caution:

$$\left.\begin{array}{l} R = \frac{W\ell}{4\alpha} \left[\frac{-1+6\alpha-6\alpha^2-\alpha^3}{3-4\alpha}\right] \\ M_s = \frac{W\ell}{8} \cdot 2 \left[\frac{1-6\alpha+12\alpha^2-7\alpha^3}{3-4\alpha}\right] \\ M_c = \frac{W\ell}{8} \left[\frac{1-4\alpha+4\alpha^2-2\alpha^3}{3-4\alpha}\right]. \end{array}\right\} \quad (6)$$

The value of R must be positive; the solution of the cubic equation gives $\alpha = 0.2142$ as the condition for lift off from the walls, and the results of equations (6) are valid only for α greater than this value.

16. It is of interest that there is no value of α in the range $0.2142 < \alpha < 0.5$ for which the values of M_s and M_c become equal. Instead, the value of M_c decreases as α increases, eventually changing sign, while the value of M_s reaches a minimum for $\alpha = 0.3500$, for which value

$$[M_s]_{\min} = (0.0873)\frac{W\ell}{8}. \quad (7)$$

According, therefore, to this elastic analysis, equation (7) results from the optimum arrangement of the brace supports if the moments in the roof beam are to be as small as possible.

17. Figure 8(a) displays the results of equations (4) and (6), the values of the bending moments being plotted against the value of α. At $\alpha = 0.2071$ the values of M_s and M_c are equal, as has been noted, equation (3), and the ends of the roof beam are clear of the supporting walls.

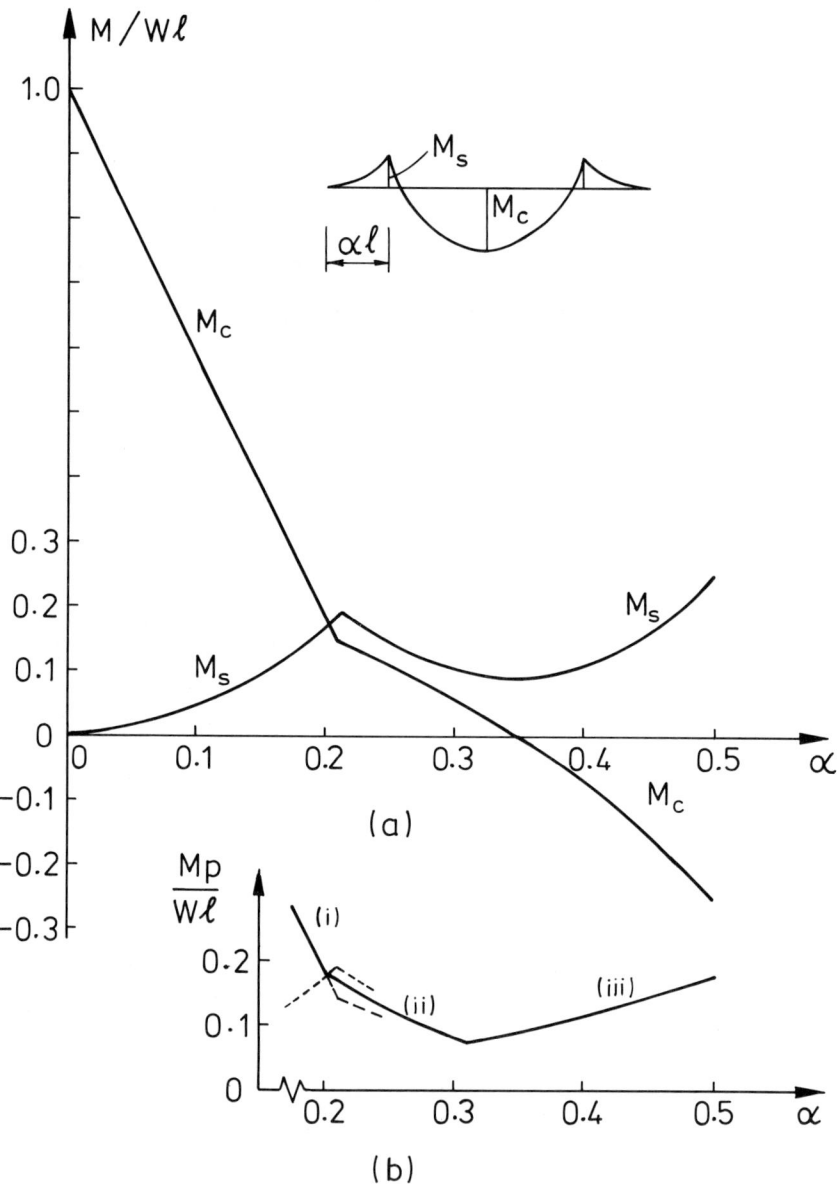

Fig. 8 Beam on four supports (a) "elastic" solutions
(b) "plastic" solutions

This "undergraduate optimum" design does not depend on the doubtful assumptions that have to be made for the elastic analysis, since the system is statically determinate. If the elastic assumptions are valid, the ends of the beam touch down on the walls for $\alpha = 0.2142$, and the "best" elastic design, equation (7), occurs for $\alpha = 0.3500$. Some corresponding typical bending-moment diagrams are sketched in Fig. 9.

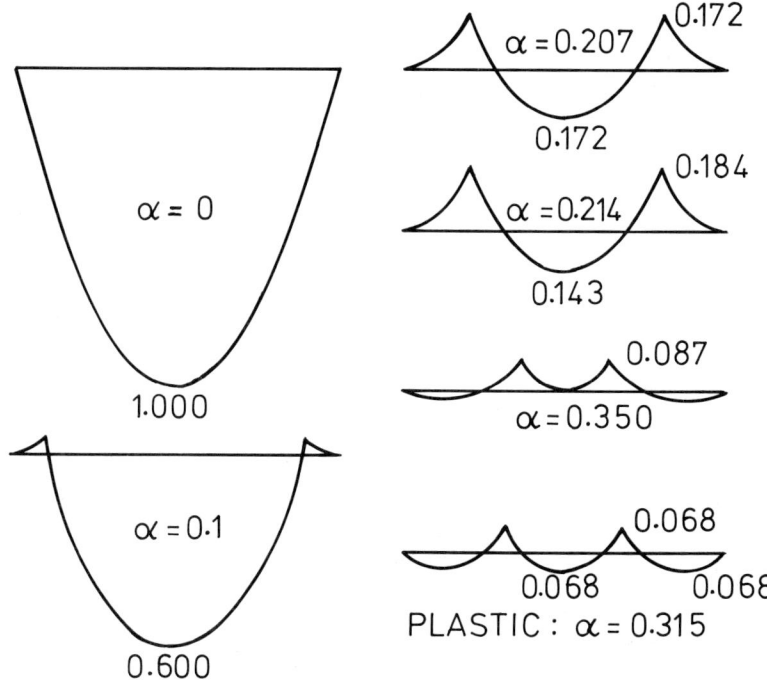

Fig. 9 Bending-moment diagrams for different positions of support braces. Cardinal values of bending moment $\times W\ell/8$.

18. It may be noted that in the range $0.3574 < \alpha < 0.4264$ the value of M_s does not correspond to the largest bending moment in the beam. Figure 10 sketches the bending-moment diagram for the end span; the value of $[M]_{\max}$ is given by

$$[M]_{\max} = \frac{W\ell}{8}\left[1 - \frac{2M_s}{\alpha W\ell}\right]^2, \qquad (8)$$

where the value of M_s is the known function of α, equation (6).

A "PLASTIC ANALYSIS"

19. Wood is not strictly "plastic", but has nevertheless sufficient ductility to make it an efficient structural material (unlike glass or cast iron), able to force a given construction into a "load-sharing" mode. A plastic solution to the present problem has at least the applicability and relevance of the elastic solution outlined above.

20. Figure 11 shows the three possible modes of collapse of the roof beam, depending on the value of α.

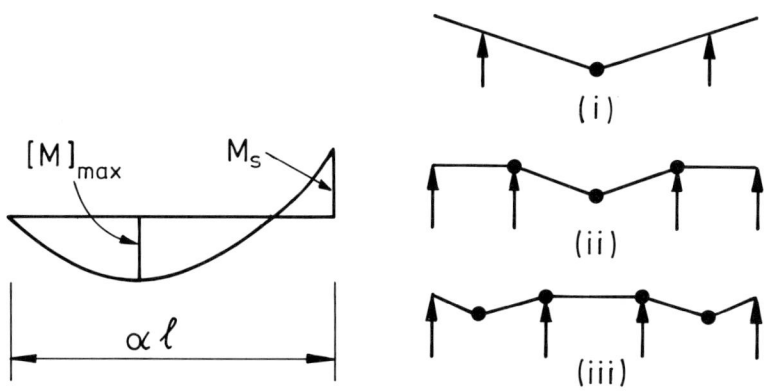

Fig. 10 Elastic analysis   Fig. 11 Plastic modes of collapse

For α small ($\alpha < 0.2071$) mode (i) will occur; this is of course the statically-determinate lift-off case, for which $M_c > M_s$. For $\alpha > 0.2071$ the mode switches to that shown in Fig. 11(ii) and finally, as α is increased further, mode (iii) occurs. The maximum design moments (M_p) from this plastic analysis are

$$\left. \begin{array}{ll} \text{(i)} \quad & 0 < \alpha < 0.2071 \quad M_p = (1 - 4\alpha)\frac{W\ell}{8} \\[1em] \text{(ii)} \quad 0.2071 < \alpha < 0.3153 \quad & M_p = \tfrac{1}{2}(1 - 2\alpha)^2 \frac{W\ell}{8} \\[1em] \text{(iii)} \quad 0.3153 < \alpha < 0.5 \quad & M_p = (0.6863\alpha^2)\frac{W\ell}{8}. \end{array} \right\} \quad (9)$$

The value $\alpha = 0.3153$ results from equating the second and third expressions for M_p.

21. Equations (9) for the "plastic" design of the roof beam are plotted in Fig. 8(b), cf. Fig. 8(a). For $\alpha < 0.2071$ the elastic and plastic designs are of course identical for the statically-determinate beam; for $\alpha > 0.2071$ the plastic design gives a value of bending moment M_p always less than the largest elastic bending moment for the same value of

α. The optimum plastic design occurs at the switch between modes (ii) and (iii) in Fig. 11, for which $\alpha = 0.3153$. The corresponding value of bending moment is

$$[M_p]_{\min} = (0.0682)\frac{W\ell}{8}, \tag{10}$$

cf. equation (7), and the bending-moment diagram is sketched in Fig. 9.

THE DORMITORY ROOF

22. It is a remarkable fact that the braces in Fig. 1 are positioned within the span to provide support at a value of α roughly equal to 0.21. It will be recalled that at $\alpha = 0.2071$ the central and support bending moments are equal, equation (3), while at $\alpha = 0.2142$ the side walls start to provide support for the roof beams. Thus the braces ensure that the wall plates are largely relieved of direct bearing, the roof loads being transmitted down the braces to the feet of the wall posts. Moreover, the idealised roof system is statically determinate for $\alpha < 0.2142$, so that some of the uncertainties of both an elastic and a plastic analysis are allayed.

23. It was noted that the bending stress in a roof beam (of the nominal dimensions quoted), had it been simply supported from the walls, would have been 9.9 N/mm^2. The braces reduce the stress to a value corresponding to equation (3), namely 1.7 N/mm^2.

24. (As a matter of record, had large, and possibly uneconomic, braces been used at about one-third span, then the absolute minimum stress corresponding to equation (10) would have been 0.7 N/mm^2.)

25. Thus the braces to the roof beams are positioned to ensure a good design; moreover the values of maximum stress, at less than 2 N/mm^2, are appropriate for a timber structure required to last through the centuries.

26. The Author is grateful to Mr Ian Curry, Architect to the Fabric of Durham Cathedral, for copies of drawings and for much detailed information.

The maintenance problem

I. HUME, DIC, Dipl Cons AA, CEng, MIStructE, Chief Engineer, Conservation Engineering Branch, English Heritage

SYNOPSIS. Although maintenance is clearly vital to the future life of a cathedral it is all too easy to let programmes slip and it is often difficult to raise funds for such undramatic things as gutter clearing and ongoing repairs. The value of proper and regular maintenance is discussed.
   What happens when things go wrong. The philosophy of repairs as described in the maxim "minimum intervention" as advocated by English Heritage is outlined.
   One way of safely adhering to the minimum intervention approach is the accurate monitoring of structures for movement and simple methods for this are examined.

MAINTENANCE.
   1. The raising of funds for maintenance is not easy yet on-going maintenance is probably the most important feature in the long term survival of any building. The work of maintenance is so very often the work of the mundane. It is gutter clearance. It is painting. It is keeping gutters and lead roofs in good order. It is replacing slates when they slip. It is mending broken windows. It is regular treatment to stop insect attack. It is pointing of masonry and it is generally keeping things clean.
   2. Good maintenance is not noticed until something goes wrong. A blocked gutter is not noticed until water begins entering the building. Blocked downpipes are not noticed until they fracture and pour water on masonry. Leadwork with pinholes is not noticed until the roof leaks. Very often such leaks are not spotted immediately because the water seeps into a void, which is not accessible, or it enters into a part of the building which is not inspected very often.

3. Proper maintenance is better accomplished when access is easy and safe. One of the roots of the problem of inadequate maintenance is that it is difficult to get to certain areas such as gutters and once there working conditions are often unsafe. It is the obligation of those responsible for building maintenance to ensure that areas which need regular maintenance are easily and safely accessible. When major repairs are in hand, future access should be a consideration.

INSPECTION.
4. Hand in hand with maintenance must go inspection, an equally mundane exercise and one which, if done thoroughly and regularly, can cost considerable sums of money. However if problems are discovered at early an stage vast sums of money can be saved and, when compared to the possible savings, inspection costs pale into insignificance. The importance of regular detailed inspections and of a regular programme of maintenance cannot be over-emphasised. Maintenance and inspections must be viewed as investments for the future.

5. Unfortunately, for various reasons, inspection programmes get delayed and maintenance is overlooked. Invariably this results in some form of decay of the basic structure of the building and damage to important finishes follows. It results in unplanned work and unforeseen expenditure. Things have gone wrong, funds have to be raised, major works are needed and there is much disruption. Proper maintenance and regular inspections may well have prevented these problems.

MINIMUM INTERVENTION.
6. When dealing with repairs to buildings, the viewpoint of the conservationist is one of minimum intervention and so far as possible, to conserve things as found. This viewpoint has to be tempered with thoughts of the safety of users of the building and of passers-by and of the overall structural integrity of the building. Also into the equation come matters of access and longevity of repairs.

7. Quite clearly a repair which leaves the stability of the building still threatened or one which results in the building being patched up for a year or two is not an acceptable form of repair even though it may conserve something as found. Equally, repairs which demand expensive access and/or which are in exposed locations may need to be treated more rigorously than other, more normal, repairs.

8. There are various ways of making repairs to structural elements. They can be improved in some

way, for example adding steel brackets to a timber roof. They can be repaired in a traditional fashion by replacing like-with-like material, for example, scarfing new ends to decayed rafters. In certain circumstances resin repairs may be considered.
Total replacement of elements of the building is to be avoided wherever possible and facsimiles are most undesirable of all. Pointing must be to the highest standard and all materials must have a long life.

9. When repairs are being considered, it is important, to establish, by research into the documentation, what was done to the building in the past. It is also necessary to check with the evidence that is presented by the building itself that the written records match the physical evidence. Very often a previous repair may affect subsequent developments and historical research is essential.

10. It must also be ensured that repairs are necessary. Some signs of distress which are perceived as real problems may not be problems at all; they may merely be signs of some past, but now stable, problem.

STRUCTURAL MONITORING.

11. It must be said that many buildings suffer various forms of structural distress (such as distortion of walls and fractures) for reasons other than lack of maintenance. Ground movements, seasonal changes and structural inadequacies are typical reasons. These movements may be current, in which case some action may need to be taken, or they may be past movements which have either ceased permanently or which are only dormant. In each of these situations structural monitoring has a role to play.

12. Accurate structural monitoring has a number of advantages:
   (a) It can be of great help in the correct diagnosis of structural ills; this means that effective and sympathetic treatment is more likely to be implemented. Before remedial works are commenced, a careful survey of cracks and other damage should be made and accurate monitoring commenced to learn more about the magnitude and direction of the movements. Such surveys and monitoring procedures can often avoid costly but incorrect solutions. For example, the building will not be underpinned when it is the roof which is at fault.
   (b) It can be a source of reassurance and an invaluable aid in convincing others that the decision to take no action to remedy an

apparent problem was correct because the building, having once moved, is now stable. Deciding to do nothing is often much more difficult than spending vast sums of money on remedial schemes which may not be necessary.
(c) It can prove that cracks which have been repaired but which have reappeared are due to seasonal and/or climatic variations and are therefore subject to opening and closing cycles rather than progressively opening. In such cases it is often not necessary to carry out any further remedial works; sometimes remedial works in these situations can be counterproductive.

13. It is often wise to establish a system to monitor movements and to keep a close watch for some considerable length of time, possibly over a period of years, before coming to a definite conclusion.

14. Many of the more simple, but nevertheless accurate, methods of monitoring available utilise equipment which is brought to the site for the monitoring exercise and then taken away for use elsewhere. The targets and stations used for these techniques are of minimal cost and are generally unobtrusive.

IS THAT CRACK GETTING WIDER?
15. It seems to be a basic fact of life that cracks appear to get wider the longer they are studied. This may be due to the crack getting dirtier and therefore more noticeable, it may be psychological or it may actually be a fact that crack is widening due to movement.

Figure 1. The Demec strain gauge in use across a crack.

16. Movements of cracks in structures can be monitored using a demountable strain gauge (the Demec) together with suitable locating points fitted to the structure adjacent to the cracks.

17. The 200 mm long gauge is shown in use in figure 1 but the manufacturers produce a wide range of differing lengths.

18. The locating points are 6 mm diameter stainless steel discs that have a small hole drilled at the centre for accurate positioning of the conical points of the gauge. These steel discs are fixed to the structure with glue. An alternative method of fixing locating points is to drill 5 mm diameter holes 24 mm deep into the structure and to insert into this a hammer-in fixing consisting of a flanged expansion sleeve and a nail. The nail is driven flush with the surface and drilled with a BS1 centre drill in order to receive the conical gauge point. These locating points are shown in Figure 2.

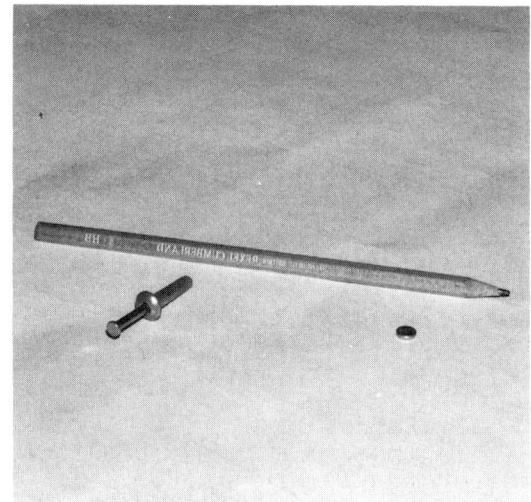

Figure 2. A Demec strain gauge point and a nail in fixing.

The type of locating point used is dependent upon the material to which the fixing is being made, the degree of exposure to the weather and the location in relation to possible vandalism. A soft wall material can make it difficult to stick the discs on and therefore the drilled in fixing is advisable but conversely it can be difficult to drill into a hard material and the sticking on of discs is easier and quite satisfactory.

19. Drilling into mortar joints is to be avoided as the fixings will eventually work loose and the readings will become unreliable.

20. The typical layout of locating points in order that cracks can be monitored for both

horizontal and vertical movements will use 3 monitoring points in a triangle.

21. Using the Demec gauge, movements of 0.025 mm are easily traceable and with care, more accuracy can be achieved. The only visible signs of this form of monitoring are three small discs at each monitoring point. No particular skills are necessary for setting up and using the Demec gauge although interpretation of the results must be done with care and thought.

22. At about £350, the Demec strain gauge, although easy to use and robust, is rather expensive for occasional use but in terms of the sums necessary for even small building contracts, the sum is not large.

23. Alternatively, a good vernier gauge (purchased from a suppliers of engineering tools) can be used to measure accurately (to at least 0.1mm) between the shanks of brass screws set in plastic plugs around a crack in the same triangular pattern. A good vernier usually incorporates a depth gauge which is a useful facility for monitoring fractures in corners of walls. Digital verniers make reading much easier.

24. The use of glass telltales cannot be recommended at all. These are susceptible to breaking from frost or vandalism, and are difficult to fix adequately. They often become detached at one end thus suggesting that by being still intact no movement has taken place. No record of progressive or climatic movement can be kept and they are unsightly, an important consideration in the field of historic building work. Mortar pats are equally ineffectual.

IS IT LEANING MORE THAN IT USE TO?

25. Changes in the out-of-plumb state of structures can be measured by one of the following methods:-
  (a) AUTOPLUMB. This is a highly sophisticated optical form of plumb bob which can be used over heights of between 2 metres and 150 metres. Because it is an optical instrument, it does not have the problems of bob swing and wind drag on the line that plumb bobs suffer from. Figure 3 shows the autoplumb. This instrument can read down to better than 0.5 mm in 10 m. A target must be attached to some high point on the structure and a small reference point is necessary at ground level. As this instrument is used at ground level only one high level "visit" need be made.

Figure 3.
The autoplumb

- (b) OPTICAL THEODOLITE. This surveying instrument can be used to check the out of vertical movements of structures to an acceptable degree of accuracy but without the need to gain access at high level demanded by the Autoplumb. There are a number of restrictions which limit the use of the Theodolite for verticality checks and which make the Autoplumb the better instrument for this purpose. As with the autoplumb, accurate and rigid targets are vital.
- (c) EDM THEODOLITE. Recent developments in Electronic Distance Measuring theodolites have now brought instruments which are sufficiently accurate for structural monitoring financially within reach. Such instruments must be set up over an accurate and rigid base station and can be made to sight on small reflective targets rather like small white bicycle reflectors fixed to the building. The EDM is designed for use with a purpose made reflector and it must be certain that it is relocated in precisely the same position each time readings are taken.

IS THERE ANY SETTLEMENT?

26. Vertical levels of structures can be recorded with a high degree of accuracy ($\pm$ 0.5 mm) by the use of a precise level with a parallel plate micrometer. Periodic checks can be made to detect movement. The levelling staff used is constructed from invar, a thermally stable alloy.

27. Again this equipment is expensive and use may be made of levels of a lesser standard. To place

any reliability on the results obtained, the level needs to be of a high standard and kept in good order with frequent checks on its adjustment. Even so an accuracy of better than $\pm$ 5 mm should not be expected.

28. It is vital to ensure that the staff is returned to exactly the same point each time monitoring levels are taken. The "left hand side of the door step" is clearly not good enough as dirt may accumulate and the step may even be lifted and reset.

IS THERE A CHANGE IN DISTANCE?

29. Horizontal changes in distance can be detected by accurate measurements with a steel tape held under constant tension by use of a tape extensometer. Rigid eyes must be fixed in the structure to enable the tape extensometer to be attached and tensioned properly. It is possible to use purpose made demountable eyes which can be accurately repositioned each time a set of readings is taken. Adjustments to the readings have to be made to allow for temperature variations.

30. It is often necessary to employ a number of different techniques in order to maintain a thorough check on any structural or ground movements. It will be wise to consider the results obtained over a period of several months (perhaps even years) in the light of experience gained from work of a similar nature and taking account of matters such as temperature, rainfall, soil conditions, state of structure, etc. before a conclusion is reached.

31. The simple but accurate monitoring of structural and ground movements is a powerful but much underrated weapon for use in the conservation of the built environment. It can be both straightforward and relatively economic as well as allowing buildings to continue in use without major structural intervention. It can also be expensive but very often a considerable amount of information can be obtained for a relatively small amount of money.

CONCLUSION.

32. Maintenance, inspections and monitoring are mundane and unspectacular but nevertheless vital for ensuring a sound future for our built heritage. They must be treated as such in spite of the difficulty of raising funds and keeping programmes on target. They usually result in considerable financial savings by avoiding massive repair schemes.

Maintenance management

S. D. STEVENS, BSc, PhD, ARICS, Chief Quantity Surveyor, English Heritage

SYNOPSIS. Maintenance should be managed. Whilst unforeseen failures do occur from time-to-time - leading to emergency repair, a proper maintenance policy will ensure that, provided a building is soundly constructed, it can have a useful life for literally hundreds of years. An essential part of establishing a maintenance management regime is the estimation, planning and monitoring of expenditure.

MANAGEMENT
1. All maintenance should be managed. To an extent it is. Albeit that, at the extreme, there is panic management - "the roof's fallen in what should we do?" Just as we need a policy for servicing our cars at regular intervals so we need a policy for maintaining our buildings. Both make common sense in the long run. But how do we compile a policy and what does it mean? Essentially, a maintenance policy is a control framework designed to keep a building in good condition - it comprises the following elements:-

Planning
Procuring
Monitoring
Feedback

PLANNING
2. This consists of two parts:-

(a) Planning pre-determined work (Cyclical maintenance) which will always be required at uniform intervals, eg cleaning gutters, painting windows
(b) Inspections at appropriate intervals leading to the identification of items of work necessary to conserve the building eg replacing stonework (Planned maintenance)

MAINTENANCE MANAGEMENT

3. The content of the programme of maintenance will not be the same from building to building nor will it necessarily repeat from year to year for the same building. For example, when dealing with a building that is in a deteriorated condition it will be necessary to instigate a programme of work to bring it up to a satisfactory state. Thereafter, an ongoing inspection and maintenance programme is required which will keep the building in its improved state.

4. The items of maintenance can be easily depicted on a simple chart showing which activities are required and when they are planned to occur. There should be an annual plan showing work each year on a rolling programme say for five years with projected broad outline of work and annual expenditure for years six to ten.

5. Items of cyclical maintenance are fairly easy to define, eg gutters need to be cleared annually in autumn. Obviously those who live and/or work in a building will notice obvious items that need attention and affect them particularly but they will invariably grow to live with the creeping deterioration that occurs and not notice the first signs of more serious problems. There is therefore a need for a full "health check" to be carried out periodically by a professional. When dealing with a large and complex building one must balance the effort and cost of such an in-depth exercise with the benefits.

6. A general rule of thumb is to carry out a full inspection every five years - a quinquennial inspection. Such an inspection should comprise a detailed look at all the elements of a building and should result in a written report including recommendations for work to be carried out with estimated costs where possible. It should also monitor and report upon work carried out during the previous five year period. It therefore becomes part of an ongoing programme of review, monitoring and feedback and a means of highlighting planned maintenance projects before significant deterioration can occur.

7. There is, of course, another maintenance heading - "emergency" maintenance. Whilst this may genuinely be an emergency reaction to an event, eg vandalism, so often emergency failure is due to the fact that cyclical or planned maintenance has not been carried out. A favourite example of this is blocked or cracked rainwater downpipes leading to ingress of water to a building with resultant damage. By investing relatively small amounts of money in cyclical maintenance, therefore, much larger sums can be saved in the long term.

8. Of course, it is easy to talk about spending money. But money is not always so easy to find. The Government and English Heritage

have long recognised the ongoing burden of repairing England's cathedrals. This resulted in the launch of the Cathedral Repair Grant Scheme in 1991. A total grant of £11.5m was made available over the three years 1991-94. In 1991/2 35 cathedrals shared £2m with grants ranging from £5,000 to £235,000.

9. I should, however, stress that such grants are available for repair but not for routine maintenance.

10. How can we plan and manage expenditure? Once the programmes of work have been compiled estimates of the likely cost of those works can be produced, enabling the production of annual budgets and projected long-term expenditure.

11. There are, however, no easy shortcuts when producing such estimates. We need to analyse every item. This is made easier when we have historic data available of the cost of previous similar work at the same location. From this we can extract costs and rates from which to estimate. Otherwise we need to look to other sources. As far as the cathedrals are concerned some procure works differently from others - they may have their own directly-employed labour, for example. So we need to look not only at the work but how it is procured.

12. Having produced our estimates of expenditure we can then analyse whether re-allocation of tasks would lead to a more favourable expenditure profile for the Client. For example, it may be that a number of items fall due in a particular year leading to a disproportionately high spend. It may be possible to re-schedule those tasks, by bringing them forward or delaying them, to produce a more even expenditure from year to year. To do this we need to analyse any additional costs involved in delaying a particular repair and any "knock-on" effects that the delay may have. Whilst it may be attractive to delay expenditure from one year to another this could result in a much higher spend overall.

13. Once the programme is determined we can consider how to procure the work.

PROCURING

14. A proper contractual and/or control framework is the key to ensuring proper value for money particularly where the extent of work cannot be fully determined in advance.

15. The documentation used to procure maintenance work particularly when dealing with historic fabric is, of course, extremely important. It should include:

Detailed drawings
Contract conditions
Specification
Schedule of Works
Schedule of Rates
Schedule of Daywork costs

16. Above all, it should be precise, clear and unambiguous. Cyclical maintenance might best be procured through a "term" contract whereby works can be ordered as and when required during a pre-defined period or term. Planned maintenance will normally be the subject of a one-off contract.

17. Whatever contractual approach is chosen, proper specification is essential based upon proper investigation and research. It is far better to expend monies on opening up and investigation work than to find that work has commenced on inaccurate or incorrect information.

MONITORING AND FEEDBACK

18. Monitoring and feedback should be an essential part of the management process and should be encapsulated in the systems which are adopted.

This takes two forms:-

(a) Ensuring that the work is carried out as programmed and that the intervals for cyclical maintenance items are about right
(b) Monitoring the performance of repairs to ensure that the correct repair has, in fact, been carried out.

19. The first is a management task which should be undertaken periodically during the year, say quarterly. The second is part of the technical inspection that should be carried out every five years. Such recording is essential in monitoring the performance of repairs and thus ensuring that mistakes are not repeated.

20. All structures, if left to their own devices, will revert eventually to ruin - ashes to ashes, dust to dust. If we do not interfere with the land it will revert to forest. Maintenance is the art of economically prolonging the useful life of a building for our own enjoyment and that of future generations.